By Brian Hare and Vanessa Woods

The Genius of Dogs

Survival of the Friendliest

Puppy Kindergarten

Puppy Kindergarten

Puppy Kindergarten

The New Science of
Raising a Great Dog

Brian Hare
and
Vanessa Woods

RANDOM HOUSE
NEW YORK

Published in the United States by Random House, an imprint and division of Penguin Random House LLC, New York.

RANDOM HOUSE and the HOUSE colophon are registered trademarks of Penguin Random House LLC.

Hardback ISBN 9780593231326
Ebook ISBN 9780593231333

Printed in the United States of America on acid-free paper

randomhousebooks.com

2 4 6 8 9 7 5 3 1

First Edition

Book design by Susan Turner

For Congo,
a truly great dog

CONTENTS

Congo

The principal of the Puppy Kindergarten starts each day with a long stretch. Congo's giant paws reach over the edge of his XXL bed, his head arches upward toward his tail. He yawns, rolls over, and gets up. Downstairs, he might snooze in front of the fire until his humans are ready. Then he climbs into the backseat of the car while he is chauffeured to work.

Congo always takes the same route from the parking lot to the Puppy Kindergarten—he walks past the camel statue, pauses, and takes in the sweeping view of the puppy park where his students are out for recess.

There has never been a celebrity more adored by their fans. Just the sight of him sends the puppies into a tail-wagging frenzy. They climb over one another to get to the corner of the fence where he will make first contact, their little paws stepping on wrinkled faces and pinning floppy ears to the ground.

Congo pretends not to notice. He takes his time, sniffing each side of the path. He checks in with the humans first and allows one of them to take his leash.

"Good morning, Congo!"

"They are ready for you, Congo!"

Then, taking a deep breath and drawing himself up to his full three feet, eight inches, Congo goes to work.

Everyone who meets Congo always says, "What a great dog."

Congo does not bark unless you ask him to and has no issue being quiet upon further request. He does not dig in the yard. He does not chase squirrels, cats, or other tempting targets. He greets strange dogs in a friendly, polite manner. He greets new people in the same way. He can as easily stroll beside you on a loose leash as accompany you on a five-mile run. If you are having a crazy day and can't take him for a walk, he will happily snooze on his bed. He is kind with the elderly and patient with children. His skill set is impressive. He is strong enough to pull a wheelchair but gentle enough to pick up a quarter you have dropped on the floor and place it in your hand. He can turn lights on and off, tug open doors, and load your laundry.

But more important than the tasks he can do is the comfort he provides. If you have had a hard day, you can ask him to lay his head on your lap. If you need a hug, he will step up carefully onto your knees and lay his head on your shoulder. Congo is a retired service dog, although now that he is the principal of the Puppy Kindergarten, he is not *really* retired. He is a test pilot for our research, helps raise the puppies, and, more than anything, has shown us what service dogs have to teach us about how to raise great family dogs.

Living with a great dog is one of the true joys in life. No matter how disappointed they might be that the morning walk wasn't twice as long or that their secret hopes of a surprise second dinner did not materialize, they are always happy to see us when we walk through the door. They are eternally optimistic—always sure that the next super-awesome-great thing is right around the corner. They bring us outdoors into the fresh air and encourage us to stop and explore places we would never go otherwise. They are great listeners. And most important, they bear witness to our lives, to every triumph and tragedy; they are a constant loving presence from the day they arrive until the day they leave us.

Almost any puppy, if loved, will show tremendous devotion to the people who raise them. And every dog is unique:

Anyone who has had more than one dog knows this. We love dogs, but we fall *in love* with their funny quirks and idiosyncrasies. There are dogs who can unlatch gates, and dogs who love riding in the car. There are small dogs who think they are apex predators and giant dogs who are afraid of their reflections. Some dogs love to play with toys and others just want to be near you.

But sometimes there is a mismatch between a dog and a certain lifestyle. A dog with boundless energy will struggle in an apartment with owners who work long hours. A dog who frequently barks may bother the neighbors. Sometimes, our expectations are too high. Puppies can get sick, chew furniture, and eat poop. Puppies need to be bathed, trained, and walked in all weather. They can pull on their leashes, and if it's raining, they can refuse to walk or go to the bathroom. They have preferences and motivations that do not always match ours.

In the worst cases, the mismatch is so serious that the bond is broken. Aggression toward people, especially children, separation anxiety, destructive chewing, soiling, and incessant barking are some of the behaviors that can ruin the relationship between a dog and their family.

Everyone who brings home a puppy wants to raise a great dog. What does it take? What do we need to know about how dogs develop to make sure we are the best puppy parents possible? These are some of the most complex—and most important—questions we can ask as dog lovers. We wanted to use what we have learned from our research with service dogs to write a book addressing these questions.

Dog Lovers to Dog Scientists

Our path to becoming puppy experts was indirect. We are two dog lovers who have lived with dogs our whole lives. We also happen to be scientists fascinated by what animals can teach us about cognition. From the Congo Basin to Siberia, we have challenged a variety of animals—from our closest primate relatives, bonobos and chimpanzees, to more distant relatives like wolves, coyotes, and foxes—with puzzles, games, and tests to uncover what they understand, and, in some cases, don't understand, about the world.

We base our tests on experimental games used with infants and young children. We also work with infant and adolescent animals to understand how their cognition develops compared to our own. This means we have spent years studying how different baby animals grow up. Often, we have found that animals are more like us than anyone suspected. Our experimental work has helped show that we share the world with other sophisticated beings with rich mental lives.[1]

One of our main findings is that dogs as a species have a social genius for cooperating and communicating with humans. This ability was enhanced by domestication.[2, 3] But some dogs have more of this social intelligence than others— even as puppies.[4] It was when we saw these meaningful differences between individual dogs that we decided to apply our scientific discoveries to the real-world problem of raising great dogs—both for service and for families.

Congo also took a circuitous path to the Duke Puppy

Kindergarten. Before coming to live with us, he was a service dog with Canine Companions, the largest service dog provider in the United States. Canine Companions breeds, raises, and trains dogs to assist people with a range of physical and developmental disabilities. Donations allow Canine Companions, a nonprofit organization, to provide each client with a service dog free of charge.*

For over a decade, we have been helping Canine Companions solve one of their biggest problems: How can they raise more of these great service dogs? Millions of people with disabilities could benefit from a dog like Congo, but there are simply not enough service dogs. Professional service dog training takes years and requires rigorous certification tests. It is harder to graduate as a Canine Companions dog than it is to graduate from the best colleges.† Each dog requires years of

---

* We work with many wonderful service dog organizations, including North Carolina–based Ears Eyes Nose and Paws. However, Canine Companions is our main partner. Founded in 1975, Canine Companions is one of the largest service dog organizations in the world. At no cost to their clients, they have placed thousands of dogs with people all across the United States. For over thirty-five years, Canine Companions has bred their own dogs, a combination of Labradors and golden retrievers.

Canine Companions breeds around eight hundred puppies a year. Litters are reared by their mothers and attentive staff. At eight weeks of age, each puppy is placed in the home of a volunteer puppy raiser for the next fourteen to eighteen months. Socialization and training of the puppies begins almost immediately; they are exposed to people and places while learning skills appropriate to their age. Every puppy receives regular physical checkups, as well as monthly assessments to confirm they are on track. Then, at around eighteen months old, each dog returns to one of Canine Companions' six campuses located regionally across the United States for rigorous professional training.

† Unfortunately, the short supply of professionally trained dogs has created an opportunity for fake service dog trainers who take advantage

investment, both in time and money, and only half will graduate. How can we have more dogs help more people? At the
Duke Canine Cognition Center, we realized that the way
each dog solves problems could give us a powerful new way to
predict earlier, and with more certainty, which dogs would go
on to be successful service animals. We did not initially set
out to learn *how* to raise great dogs. But as we helped Canine
Companions, we discovered the lessons we were learning
about service dogs could be applied to raising dogs at home.

The most important lesson we learned is that each dog
is an individual. Individuality is produced when lots of traits
vary independently. Someone who is tall does not necessarily have perfect pitch. A skilled mathematician might be a
terrible writer. This type of variability also exists in dogs,
meaning that some dogs excel at certain types of training but
not others. For example, the same dog might be highly trainable for a job helping a veteran with PTSD but not as trainable for a job helping someone with a physical disability. Or
a dog in training to help someone with disabilities might be
better suited to learn how to detect contraband at the air-
port. It is often hard to tell which job, if any, will suit a dog
best until after they begin training. This uncertainty costs
time and effort and reduces the number of dogs who graduate from professional training programs.

---

of those desperate to find a dog to help their child, parent, or loved
one. These unscrupulous people charge hundreds or even thousands
of dollars for a dog that has little or no training. One solution is to
raise awareness and enact laws to protect consumers from fraud. For
example, airlines have more restrictive definitions of what is allowable,
and bills are being passed in some states to require basic standards for
anyone claiming to be a service dog trainer.

We decided to turn the problem of dogs' individuality into an asset. If we could find a way to measure behavioral variation between dogs, we might be able to use certain traits to predict which individuals would succeed at different types of jobs before they even began training. This approach has previously been used with some success. Individuality in temperament, including emotional reactivity and motivation, are already measured to predict training success.[5]

If a dog is fearful or stressed when confronting new people and new places, they will not, even with training, be able to help someone else navigate these environments. If a dog is not motivated by rewards, like toys or treats, they will not be motivated to solve problems when they are working. Service dog trainers look for a calm, easygoing temperament. Together with positive training, temperament assessments have become standard tools. But while they have helped improve service training outcomes, there are still not enough professionally trained dogs.

This is where our cognitive-based approach comes in. Watching a service dog train, you can see they are constantly using their minds to solve a range of problems. Service dogs need self-control to keep them on track when they want to play with other dogs or chase a squirrel instead of working. They rely on their memories when they have to remember skills, past situations, and different people and places. They need to understand the gestures, facial expressions, and vocal signals humans use to make requests. They have to infer what people want or do not want or where to find something, even when they can't see the object in question. Service dogs also must pay attention and respond to the needs of the person they are helping. Inhibiting impulses, recalling, understanding, recognizing, and thinking about the thoughts of others—all these skills require cognition.

Cognition is clearly critical to service dog success, but no one had ever done a large-scale study of individual cognitive differences in any working animal, aside from humans. There were no studies on memory in service dogs—even though remembering dozens of skills is critical to the job. Nor was there any research on how well service dogs understood gestures, even though, again, this is central to their work.

We suspected that cognition was the missing piece of the puzzle. Temperament motivates an animal to solve a problem, but their cognition allows them to solve it. A trainer can teach a dog to display a behavior for a reward, but a trainer cannot teach a dog to remember things for long periods of time or imagine solutions to new challenges. Neither temperament nor training allows dogs to flexibly solve problems they have never seen before. We realized that identifying particular cognitive skills could help us predict which dogs would be most likely to succeed as service dogs.[6]

Our holy grail became the development of a standardized set of tests we could use to assess a group of untrained puppies and determine which of them have the most potential for training success.[7] With a tool like that, we could not only identify remarkable dogs while they were still puppies, we could also accurately predict the career each dog would be most likely to succeed in. Were they better suited to help an autistic child sleep through the night or aid someone who is physically impaired with everyday tasks? Were they good at detecting explosives, diseases, or drugs? If we were able

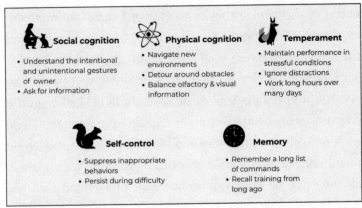

Examples of different cognitive skills and how they might be important to service dogs

to develop each dog's particular gift, we could reduce cost, increase the supply of professionally trained dogs, and, most importantly, make more dogs and people happier.

The only way to find out if puppy tests like this might help was to answer a few fundamental questions: What are the cognitive skills that puppies rely on during training and when do they first develop? What experiences do puppies need to make sure they are ready to learn? If a puppy outperforms other puppies, will they still outperform other dogs once they have grown up? To do this research we needed to raise and test *a lot* of puppies. This is why we founded the Duke Puppy Kindergarten.

Professional Dog to Family Dog

You might be wondering what all this has to do with you and your puppy.

A cultural shift has occurred in the way we live with our

dogs, and many of the traits we look for in service dogs also happen to map onto what we want in our family dogs. Dogs come with us on vacations, pose for pictures on holiday cards, and are present at our most important and meaningful events. This was not always the case. Only a hundred years ago, pet dogs were valued for the work they could do— hunting, shepherding, or guarding. Dogs needed to be high energy and aggressively defend property against strange people and animals. Many of us can still remember a time when dogs roamed around the suburbs, slept in outdoor dog houses, and often had fleas. If a dog ate something gross then had diarrhea for days, decided their life goal was to catch a car, or was less than polite to the mailman, it wasn't a big deal. But these characteristics no longer fit with the lives we imagine for our dogs. Our culture has shifted, and we have elevated the status of dogs from pets to family members, a promotion that brings added responsibilities. So, our dogs today have a lot to learn from service dogs: how to walk nicely on a leash, sit peacefully with us at an outdoor restaurant, and greet new friends calmly and politely. And in reality, service dog puppies are raised a lot like your puppy. They go home to a family at around eight weeks old and are treated pretty much like the average family dog, only with a few more rules, such as not being allowed to jump on furniture or play too rough with other dogs. Otherwise, the early experiences of your pet puppy and service dog puppies should be fairly similar. They will grow up with you and the family you live with. They will visit new places, like a veterinary clinic with strange smells, a busy sidewalk with loud traffic, or a park full of other dogs. Most likely, they will grow up without similarly aged puppies in the house even if you have another dog. They will meet a variety of people of all ages and walks of life.

When we considered the similarities between what service dogs have been for decades and what we hope our pet dogs will be, we realized that what we learned about service dogs could help everyone raise a great dog. Our team has raised dozens of puppies and collected data on hundreds more to better understand how to get the best outcomes for both service and pet dogs.

We begin *Puppy Kindergarten* by explaining the origins of our new understanding of dog development. We compare brain development in dogs to humans and other animals to determine at what age we should study the development of their mental abilities. We also share breakthroughs we have recently made in understanding what it takes to succeed as a service dog and the role domestication played in shaping these abilities as dogs evolved.

On this foundation, we created a tool called the Dog Cognition Longitudinal Battery to uncover patterns of cognitive development. We used this tool with a group of our puppies to chart how their cognitive abilities first emerged during their final period of rapid brain growth. To understand how different rearing approaches might impact this emergence pattern, we also raised our puppies in two different ways. One group was raised in homes, similar to how you would rear your own family dog, and another group was raised at Duke University in our Puppy Kindergarten, where they received an unusual amount of early social contact.

As a result of our studies, we can now provide scientific answers to some age-old questions:

- **Selection.** Are there certain cognitive traits associated with training success? Are these traits stable over time—if a dog has a certain cognitive skill when they are young, will they still perform the same way as they get older? When do important cognitive skills come online; is it during the final period of most rapid brain development?
- **Breeding.** Can these important cognitive skills be passed down from parents to their puppies? Is it possible to identify genetic markers associated with these skills?
- **Rearing.** Is there a way to influence the development of these cognitive skills with early experiences, especially during a puppy's critical socialization window?

We find that compared to other animals, all dogs are remarkable in their ability to cooperate and communicate with humans, that dog development is more complicated than once thought, and that it is possible to make predictions about an adult dog's success in a real job in the real world based on what they were like as a puppy. While we draw on information collected from a range of different approaches, the core of our new understanding comes from our work with service dogs, and we translate what this means for the success of our own dogs at home.

The second part of the book includes hard-learned lessons from the day-to-day running of the kindergarten. The Duke Puppy Kindergarten signs up around a hundred volunteers every semester. The volunteers are all undergraduates, with varying backgrounds and interests, who turn up twice a week and devote themselves to the puppies for a two-hour shift. The volunteers wipe the sleep from puppy eyes, brush their teeth, and bathe them. They empty diaper pails and scour the grass for poop they might have missed. Most important, the volunteers take the puppies out to see the world. The puppies are welcome wherever they go: the dining hall, the library, the wellness center, and even the occasional lecture. It is the devotion of the volunteers, and the generosity of the Duke community, that has turned the entire campus into the puppies' home.

We reflect on the lessons we learned when we were forced to transfer the kindergarten into our home during the

Fearless, making off with a slipper

pandemic. We deal with the questions any puppy parent struggles with: Why won't they pee outside? Why are they chewing through the drywall? And when will they *sleep through the night*? As we recount how we struggled to get our puppies to sleep, we review what we know about their sleeping habits, whether they dream, and how their memories work. This part of the book has everything we wish we had known when we were starting out. Congo came to us when he was already an adult, so we never knew him as a puppy—but we knew every puppy we raised at the Puppy Kindergarten was supposed to *be* Congo when they grew up. Which led us to the questions we asked ourselves over and over: How do you rear a puppy to set them up for later success? Is there a way to influence the development of their cognitive skills with early experiences?

Though we outline the basic life skills and commands* your puppy will need, this is not a book focused on training. It considers the broader questions of why we raise dogs the way we do and what strategies seem to work best when you are trying to raise a great dog. There is much more to consider than training alone. So, if you have bought this book the day before your puppy arrives, and you are wondering where they will sleep, what they will eat, and what a typical day looks like, we can help (just flip to Appendix I where we have a list of supplies you need to get before they arrive). If you bought this book in a panic because you are not sure if

---

* Throughout this book, we will refer to commands as "skills," since a command implies that a puppy does not have a choice, and a skill is something you learn with practice, over time.

it is you or the puppy behaving badly, then go to Appendix II that has an example of a schedule that incorporates teaching puppy manners as well as giving them enough exercise and stimulation to wear them out by the end of the day.

After you have dealt with the immediate emergencies and are snuggled up with your puppy in a bean bag, we hope you will enjoy reading about what is happening in their funny, incredible brains—why they can't figure out the physics of a water bowl, but can read your mind the way no other species can.

Welcome to the Puppy Kindergarten. We are so glad you are here.

Puppy Kindergarten

CHAPTER 1

The Puppy Brain

Puppy Brain Growth

I f we want to understand how to predict who a puppy will grow up to be and influence the chances of them becoming a great dog, we have to first consider how their brains grow. When do puppy brains grow most rapidly and start to produce their individuality? When are their brains most plastic and influenced by experiences that affect their adult personalities? Should we start studying them when they are two or twenty weeks old?

Answering questions like these requires understanding how dogs mature compared to other animals. We know more about the brains of humans than any other species, and in many ways, puppy brains are similar to ours. Even though we tend to assume that our minds become more impressive as we get older, it is when we are young that our most extraordinary mental feats occur. In infancy, our minds are astonishingly plastic. We amass information at a rapid rate. In our first four years we learn thousands of words. The physical world begins to make sense—important concepts like gravity and the link between danger and pain. We develop morals, our emotions become more sophisticated, and we begin to think about the world and our relationship to others as we move about in it. Our psychology is shaped by who we interact with, where we spend time, and what we play with, whether it's books, baseballs, or banjos.

No matter what we do with our lives as adults, it is this period of early cognitive development that should be considered one of our greatest accomplishments.[1] Within months, we go from being helpless—unable to survive a few hours without shelter and care—to a walking, talking, culturally capable human.*

There are many ways for animals to grow up. Baby snakes slither out of their eggs so perfectly formed that they do not need a moment of parental care. Axolotl salamanders never grow up, and their bodies remain in a juvenile aquatic state for the rest of their lives. Bluehead wrasse start life as female but can change into males. The list goes on.

For most mammals, the main difference in development is how ready they are for life just after birth. Many mammals are born to run. Antelope, for example, stand up within minutes, and in the span of a few hours, they can keep up with the herd; within months, they'll be ready to strike out on their own—all useful skills when you are at the top of the

---

* Development can be described as the entire time between the fertilization of an egg to the maturation of an adult, but it is more commonly used to describe these early stages of rapid change.

predator menu. By contrast, other animals, like baby orang-utans, cling to their mother twenty-four hours a day for years.

While most mammals fall somewhere between these extremes, both dogs and humans are toward the helpless end. Mammals who need more investment from their parents tend to grow larger than average brains. Larger brains usually belong to animals with more complex behavior and more flexible problem-solving abilities. Longer periods of parental help give the maturing brain time to grow in safety and allow experience to shape growth.

Since humans and puppies are born nearly helpless, our knowledge of how the human brain develops provides insight into when puppy brains grow the fastest and are the most plastic. The comparison reveals striking similarities and conspicuous differences. The similarities help explain why puppies can immediately become part of our families while the differences can inform our expectations of what a puppy might be capable of and when.

With around eighty-six billion neurons, the adult human brain is approximately three times the size of a chimpanzee brain. Like humans, dogs have a high number of neurons in their brains compared to other carnivore species. A large domestic dog, like Congo, has over twice as many neurons as a house cat. This is true of both the number of total brain neurons and the cortical neurons, which are engaged in complex problem-solving. Congo also has more neurons, and the associated cognitive computing power, than larger carnivores like African lions or brown bears.[2] These size comparisons suggest dogs can potentially out-compute most other carnivores, but compared to humans, dogs are very limited in more sophisticated forms of cognition, such as reasoning that requires making an inference.

When compared to those of most mammals, dog brains, like human brains, are undeveloped at birth. The cortical layer of the mammalian brain consists of bumps and grooves known as "gyri" and "sulci." Folding of the outer cortical layer allows brains to pack more neurons into smaller spaces. Puppies are born with smooth brains and relatively few neurons. Puppy brains only develop folds, cortical neurons, and the resulting cognitive abilities after birth and are entirely dependent on their mothers until this brain growth occurs. This leaves newborn puppies almost as helpless at birth as human babies, and just like our own babies, parental care is critical for puppy survival.

In some ways puppies are born even more helpless than humans. At least we are born with our senses functioning. We can see, and quickly develop a preference for, the face of our mother and people who look like her. We can recognize her scent and are less likely to cry when we smell her. We immediately recognize her voice and prefer it to any other. We can even tell the difference between her language and a foreign language.* Our sense of touch is also present at birth, and every part of our body is sensitive to the physical world. In contrast, puppies are born with weak senses. New-

* But from here, everything progresses slowly. You spend your first year parsing speech from the cacophony of everyday sounds. The auditory cortex, which is complex and impressive, takes many years to finish developing.[3]

born puppies are blind, and their eyes do not open for two weeks. Although puppies can smell at birth, their olfactory cortex is not developed and their sense of smell is poor. Puppies are born with their ear canals shut; they open during the first two weeks of life. Puppies do not reliably start responding to sound until they are around twenty-five days old.[4] Hearing is less developed than vision in newborn puppies—whereas in humans it is the opposite.

The one sense that newborn puppies can rely on is touch. Shortly after birth, puppies depend mainly on body heat to find their mother's nipple. Puppies are also born with whiskers, which are specialized hairs with follicles full of nerves. They are located on a puppy's muzzle, jaw, and above the eyes. The smallest particle sends tremors down a whisker, and the nerve endings send messages to the brain. Puppies immediately begin to use their whiskers to navigate in the dark, crawl through tiny spaces, and detect the location and speed of moving objects by the airflow.

However, the other senses quickly catch up. They mature at a pace that maps onto their rapid brain development. The occipital lobe, the visual center of the brain, is the fastest developing part of the puppy brain. By day twenty-five, puppies begin to see forms and start to orient toward visual stimuli like bright light.[5] By their six-week mark, puppies can see, but it takes a few months for them to develop the full vision they will have as adults. Their olfactory bulb is more mature at two weeks and eventually grows into an incredibly complex structure—with olfactory neurons regenerating throughout adulthood.[6]

Another advantage the youngest puppy has over a human newborn is their relatively well-developed motor cortex, which is involved in the control and execution of voluntary movement. While we flop around helplessly for months, unable to stand or even sit up, puppy muscle tone develops quickly. At only a few days old they can right themselves if they are on their side, and shuffle forward to find their mother's nipple. By the third week they can sit up—then stand almost immediately after. By four weeks they can walk, and by six weeks they can right themselves when they are in danger of falling.[7]

As puppies quickly strengthen their senses and motor abilities, the rest of their brain finishes what takes our brain years to accomplish. Gyrification, or the growth of cortical folds that allow for high neuron densities, is complete by six weeks of age.[8] The full length of the corpus callosum, the part of the brain that connects the left and right hemispheres and allows them to communicate, reaches adult form at sixteen weeks. Likewise, the relative white to gray matter intensity and myelination of critical neuron networks reaches adult levels in dogs by sixteen weeks and is largely completed in the first year.[9]

This overall pattern, in which puppies grow a more neuronally dense brain than many other mammals in the first few weeks after their birth, means that puppy brains have the potential to be heavily affected by experiences. It also means a puppy's period of maximal brain plasticity takes place over a much narrower window of time than ours.

Like human brain development, puppy brain development is affected by experiences outside of the womb. Social experiences, in particular, heavily shape a dog's brain development between weaning and the appearance of adultlike brain structures at around eighteen weeks. So the final period of rapid brain growth and myelination between eight and eighteen weeks—which is, in part, why it is believed to be the critical period of socialization.

The evidence for rapid brain growth during this ten-week period is also a reminder to be patient. Puppies are not working with a full deck of cards for months after we bring them home. Puppies do not pee inside on purpose or chew on furniture to make you mad. Their brains are just slowly catching up to our expectations.

Ready for School

By eight weeks old, puppies are newly weaned, their senses have begun to develop, and they are ready to go to school. By eighteen weeks puppies' brains are largely developed and they should start showing adultlike cognitive skills as well as their individual personalities. We predicted that studying the cognition of puppies during this final period of rapid brain growth could, for the first time, pinpoint when the types of cognition that determine training success first emerge. We could then use this information to understand how social experiences might influence the expression of social abilities critical for the development of successful service dogs.

At eight weeks old, each puppy's temperament starts to

emerge. And so does their cognitive profile. So, we welcome eight-week-old Canine Companions puppies to our kinder-garten. This is the moment where we can start asking them questions and try to figure out what is happening in their extraordinary and rapidly developing minds.

CHAPTER 2

Ready for School

Human-like Social Origins

The Puppy Kindergarten is tucked away in a quiet corner of Duke University's campus, in Durham, North Carolina. We are near Duke Pond, a wetland conservation area that blooms with wildflowers from spring until fall. If you take the path by the pond and go up a long flight of stairs, you will arrive at the puppy park, a green area just large enough for half a dozen puppies to romp around in, with a fence tall enough to contain a tiger, and surrounded by a butterfly garden. Just opposite the puppy park is the redbrick Biological Sciences building. The first door leads to the three rooms of the Puppy Kindergarten—the classroom, the bedroom, and the playroom, where the class of Fall '18 is waiting for school to start.

Each semester, Canine Companions does their best to send puppies from several different litters, half males and half females. While the puppies are on campus, a hundred

Duke undergraduates volunteer to help us take care of them. It is hard to imagine how a group of puppies could be more loved by more people.

To discover when their essential cognitive skills first emerge, we host each class of puppies during their ten critical weeks of brain development. But what skills should we measure and how should we measure them? If we want to understand the origin of individuality in our dogs, predict who they will grow up to be, and influence their chances of being great dogs, we need to understand their most extraordinary cognitive ability.

Here is where it gets surprising. Puppies already have a type of social genius.

Fall '18, from left to right, Dune, Ashton, Aiden

The class of Fall '18 is more popular than any boy band: Individually, they are adorable; together, they are sensational. You can find them on campus by following the squeals of fans, who fall to their knees when Fall '18 approaches. Phones blink with Fall '18 sightings; you can almost see the clouds of emojis floating in the air after them.

The plan is to have four to seven eight-week-old puppies arrive each semester—but since Fall '18 is the first year of the kindergarten, we decided to start with three. Ashton is the front man. Shaggy russet curls fall around his shoulders. Soulful eyes tell each fan he's been waiting for them his whole life. Dune is the bad boy. He wears a permanent scowl and looks as though he hasn't slept for weeks. He is as black as a raven and his favorite trick is to murder every stuffy in the room. Aiden is the smallest member of Fall '18 and is as white as a daisy. His eyes are so big, he looks like an owl. When Aiden has the zoomies, he is so fast, he parts the air-

The puppy puddle: Aiden on top of Ashton on top of Dune

waves. But he tires easily. His favorite place to take a nap, for maximum cuteness and effect, is on top of everyone else.

Luckily, Congo is not the type to just lie around eating

bonbons all day. No one told Congo he would have to teach at the Puppy Kindergarten. He took one look at the chaos in the puppy park and decided to impose some structure.

First, the puppies must learn to play appropriately. No Canine Companions graduate is allowed to crash tackle another dog, grab the scruff of their neck in their teeth, and shake vigorously. Should the puppies try this with Congo, he picks up a toy and encourages them to play tug-of-war. If they continue to nip at him, he puts a paw on their head in a gentle reprimand, then offers the toy again.

Congo seems to understand that these puppies are not just miniature dogs. He has dog friends who are about the size of a fourteen-week-old puppy, and Congo tears around the yard with them. They wrestle like big dogs, sweeping the legs out from under one another, trying to get their mouths around one another's heads.

Congo never does this with the puppies. He is constantly self-handicapping, rolling over onto his back and letting them climb all over him and lick his mouth.

Occasionally a puppy puts their entire head in Congo's mouth, just to see what is inside. Congo waits patiently until they are satisfied, his teeth never messing up a single strand of puppy fur.

Congo is the embodiment of calm, his tail wagging a steady swish, a shining example to the puppies who wag their entire bottoms so frantically they have trouble walking.

He also leads by example. Sometimes, we take Congo on a walk with two other puppies. He demonstrates walking with a loose leash, so the puppies (who would follow him adoringly right off a cliff) learn to walk at the pace of the person. During training, Congo is quick and attentive, refusing to be distracted, despite the puppies' best attempts to get his attention.

And like any good leader, Congo knows his limits. Once he has successfully delivered his lesson for the day, or he has decided that the lesson has gone on for long enough, he goes to the door of the puppy park and barks once, waits for a volunteer to open the gate, then goes inside the kindergarten to take a well-deserved nap.

Many people have watched Congo in wonder and asked how they could get a dog like Congo to help them with their puppy. We tell them that as puppy parents we must be like Congo—or as close to him as possible—consistent and gentle with limitless patience.

Congo demonstrating a perfect sit while Aiden tries to climb in
Brian's lap and Ashton watches from under a bench

The Thoughts of Others

What do we mean when we say Congo is "teaching" the
puppies? When Congo leads by example, is he aware of the
fact that the puppies are learning from him? Does he understand that he knows things the puppies do not? Does he
know the puppies have to be watching and paying attention
to learn from his demonstrations?

Human teachers are constantly thinking about what
their students do and do not know. They're always scanning
the classroom, making sure students are paying attention.
Thinking about the thoughts of others is called "theory of
mind," because you must consider—or have a theory
about—what is in someone else's mind. It is critical to everything that makes us human.[1] Theory of mind is our greatest asset as we try to influence the minds and behaviors of

others. It is also the developmental foundation of all our cultural knowledge—including language. And it all starts with pointing.

Pointing is a curious gesture. In humans, understanding the psychology behind the act of extending the arm, hand, and finger indicates a surprising level of cognitive sophistication. Before we were nine months old, if someone pointed, we just looked at their finger. After nine months, we started to understand they were making a gesture toward something beyond their finger. We had to have an idea of what they might be thinking in order to land on what they were pointing at. We had to know the context of their gesture and think about what they might intend to communicate. Inferring the intention behind a pointing gesture is just one of the many theory-of-mind abilities that we have, but it is among the first to appear and it allows us to rapidly learn from everyone around us. As babies, we learn from adults when they point to objects, places, and people. Before our second birthday, we used this technique to learn dozens of words a day. When our parents pointed to something and said "Look!" we learned to discern whether to feel excited or scared based on the tone of their voice or facial expression. When we were confused about how to open a box or hold a toy, someone pointed to the latch that needed to be pressed or the handle that needed to be held. We succeeded because we inferred that someone's gesture was intended to help us.[2]

As our brains continued to mature, more complex inferential abilities built on these initial interpretation skills. At twelve months old, we began to point out things to others to share our discoveries with them. By sixteen months old, we would wait until someone looked at us before we pointed, because we knew they must be paying attention before they could see and understand the intention behind our gesture.

From here, our theory-of-mind abilities advanced rapidly. By four, we realized that others do not always know what we know. Before this cognitive jump, we believed everyone knows everything we know—even if they were not present to experience an event. This new understanding, that different people can have different knowledge, allowed us as toddlers to start keeping secrets. And lie. We could also teach someone something we realized they did not know. Because these theory-of-mind abilities are so crucial to the development of culture, language, and other defining traits that make us human, the scientific consensus was that even the most basic theory-of-mind abilities were unique to humans.[3]

Now we know differently. Although we do not have any scientific evidence that Congo can recognize when he knows something the puppies do not—a skill required for intentional teaching—we do have strong evidence that Congo and other dogs have another type of theory-of-mind ability. They can understand when a human intends or wants to help them—especially when they are interpreting our gestures. And adult dogs aren't special in this regard. This exceptional ability, or genius, is already present in puppies.[4]

Allow Ashton to demonstrate.

In the classroom, Ashton watches a magic show—an experiment we call the shell game. The classroom is sunny, full of new toys, games, and comfortable beds. Most important, there are always at least two humans who dispense lavish praise and an endless supply of treats.

The magic trick is an old one. In the eighteenth century, magicians used to hide a pea under one of three half walnut shells. A magician would move the shells around, while an audience member guessed where the pea would end up.

The puppy version is based on the same idea but is slightly different. We show Ashton two bowls. Then we hide the bowls behind a solid screen. We show Ashton a treat and hide it in one of the two out-of-sight bowls. Ashton sees the treat, but because of the screen, cannot see which bowl we hide it in. We remove the screen and . . . voilà! Where is the treat?

When we let Ashton or any other dog try to find the treat on their own after hiding it this way, they locate the food around 50 percent of the time. Since they only have two spots to choose from, being correct half of the time means they are guessing. As if tricked by magic, dogs searching on their own cannot outperform chance.

Now we are ready to test if Ashton can read our gesture in order to find the treat, since we know he cannot find it on his own.[*] Again, we hide the treat the exact same way, but this time we give him a clue. To help Ashton, we extend an arm and point at the bowl hiding the treat.

---

[*] Experimental situations like this, where variables have been carefully controlled for, allow us to test between different explanations. Since we know that dogs left to search for hidden treats on their own merely guess at the location, we can confirm that blocking Ashton's view of the actual hiding prevented him from seeing where it was hidden. We verify he did not hear us hide the treat, because we touched both bowls with the treat, making the same sound each time. We also know he can't smell the location of the treat—along with other research groups, we have run dozens of experiments to rule out this possibility.[5]

Ashton in the classroom. A treat has been hidden in one of the bowls, and Ashton must figure out which bowl.

Ashton doesn't hesitate. He waddles over to the correct bowl.

Good boy, Ashton.

Next is Aiden. He zooms down the mat so fast he breaks the sound barrier. He also follows the pointing gesture to find the treat. Dune is last. He is not in the mood for school. But he does love treats. He also likes to be right. So, in the end, he too follows our gesture correctly. The puppies are no longer just guessing. They recognize our pointing gesture is meant to help and they find their treats with ease.

Backyard Science

Over twenty-five years ago, Brian played this same shell game with his childhood family dog, Oreo.[6] Oreo loved play-

Aiden having a big yawn

ing fetch. He had a gift for fitting three tennis balls in his mouth at the same time. He would amble toward Brian,

his cheeks bulging like a chipmunk who ate the whole bag of nuts.

If Brian threw the three tennis balls in quick succession, Oreo saw only where the first one went. When he brought back the first ball, he would look at Brian expectantly. Brian would point to where he had thrown the second ball, and Oreo raced off in that direction to fetch it. They did it again on the third ball.

Much like what we saw take place in the classroom magic trick—where the Fall '18 puppies used a pointing gesture to find food—Oreo did not see where the ball went, so he used Brian's pointing gesture to find it.

In college, when Brian learned about theory of mind and how it develops in humans, he thought about Oreo playing fetch and wondered what else Oreo could do. So he played the shell game with Oreo—the same magic trick that we would go on to play in the Puppy Kindergarten. Brian pretended to hide a treat under two different cups, concealing its true location. Oreo knew the treat was hidden under one of the cups but not which one. Then Oreo easily followed Brian's pointing gesture to find the hidden treat. It was this simple discovery that launched two decades of research into the genius of dogs.[7]

At first, it seemed likely there was a simple explanation. But all the probable explanations, like dogs just use their sense of smell, were ruled out using different controls. In fact, dogs are able to recognize the helpful nature of some novel human gestures. They can understand if we point with our feet instead of our fingers, or if we signal the correct choice by putting a little wooden block next to the right cup. They know how to read these new gestures on their first trial even if they have never seen them before.

This cognitive ability wasn't specific to a special connection between Brian and Oreo. Dogs can understand the gestures of a stranger as well as those of their owner. Dogs can even find hidden food using the direction of a human voice.[8] And they can understand one another's gestures too: If a dog points their body toward the correct cup, other dogs can use this gesture to find the food.[9]

Simple explanations could not account for this set of experimental results. Dogs were clearly reading the helpful intentions behind our communicative gestures and passing the same tests that provided evidence for this basic theory-of-mind ability in human infants.[10]

Bonobos and chimpanzees, our two closest living primate relatives, can usually run circles around dogs when solving problems. When we tested bonobos and chimpanzees on their understanding of the physical world—for example, that solid objects make noise when they collide and cannot pass through one another—great apes succeed where dogs struggle.

But when we tried to help bonobos and chimpanzees find food by pointing at the correct location, they only guessed randomly. If we gave them practice, they could learn, but it took dozens of failed attempts. And if we switched to a new gesture, like putting a marker near the correct location, they went back to guessing. Unlike dogs, bonobos and chimpanzees do not seem to understand when gestures are intended to be helpful.[11]

This is what makes the performance of dogs so remarkable. They outperform our closest relatives with a set of skills that are thought to be critical to cultural forms of learning in young children. Our dogs can infer what we want or do not want using our gestures, in a way that bonobos and chimpanzees cannot.

It seems that humans and dogs have one type of intelligence for solving social problems and a different type of intelligence for solving problems related to the physical world.

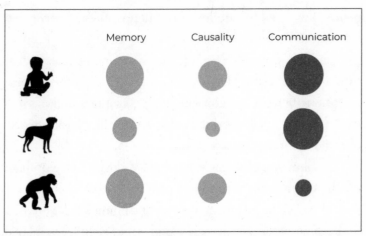

| | Memory | Causality | Communication |

Human infants and chimpanzees usually outperform dogs on a variety of cognitive measures including memory and causal reasoning. The exception is a set of games that require dogs to understand human cooperative-communicative gestures. Dogs perform similar to human infants and outperform our closest great ape relatives. Each circle represents a skill with a larger size representing more skill with a type of problem.

In other words, dogs and humans share a specialized intelligence for cooperative communication that helps them navigate their social world.[12]

In this way our dogs are the animals most socially similar to our own infants. Psychological convergence of this kind between two distantly related species is extremely rare, and the way it evolved explains why dogs are so special.

Ancestry versus Domestication

When did dogs become so humanlike? Since dogs evolved from wolves, one possibility is that dogs inherited their humanlike social skills from their wolf ancestors. Wolves and dogs share many physical and behavioral traits and are so closely related genetically that some have argued they should be considered the same species.

Wolves are cooperative hunters. They must keep within striking distance of their prey and coordinate the attack with one another. They must predict their prey's next move as well as the moves of their packmates. Wolves must be sensitive to the subtle gestures of other species, for example if their prey turns its head before changing direction, and this skill might have been generalized to understanding the subtle gestures of humans.[13] Perhaps dogs simply inherited their unusual ability to cooperate and communicate from their wolf ancestors.[14]

The only way to find out is to actually compare wolves and dogs. But adult dogs and wolves could not definitively answer the question. Adult dogs typically have so much experience with humans that it would not be a fair comparison

Wolf puppies at the Wildlife Science Center, where the testing took place. Wolves are social, cooperative pack animals. *Photo: Jennifer Bidner*

to adult wolves, who tend to be very nervous around humans. Puppies, both dog and wolf puppies, were the answer.

To make the test fair, our team worked with dozens of wolf puppies who had been hand raised by humans since a few days after they were born. When we compared the wolf and dog puppies, the wolf puppies had spent more time in the company of humans than had the dog puppies. This meant the wolves had more opportunities to learn how to communicate with humans before we started testing them.

We used the same magic trick game we played with Ashton and the Fall '18 puppies: We hid food in one of two bowls then gave the puppies a clue to where the food was. Sometimes we pointed and other times we placed a small wooden marker. If the puppies already understood human intentions, they should easily infer that the wooden marker signified the same thing as the pointing—we were trying to help them find the food.

The dog puppies were more than twice as likely as the wolf puppies to correctly follow human gestures. While the wolf puppies were slightly above chance at using a pointing gesture, the dog puppies excelled. And while the wolf puppies only guessed what the marker meant, the dog puppies could infer the meaning on their first trial. Dog puppies had no need to slowly learn an association—they just spontaneously understood this brand-new gesture.[15]

So we found no support for the idea that dogs simply inherited their unusual understanding of human gestures from wolves. Dogs also do not slowly learn how to associate human gestures with rewards. This genius of dogs comes

Wolf puppies, although adorable, cannot read human gestures like dogs can. *Photo: Jennifer Bidner*

online when their brains are still immature, when they are still barely able to see or hear. Evolution that occurred during the process of domestication has shaped the development of dogs; they are born prepared to infer our intentions after very little interaction with us. A simple understanding of human intentions is one of the first cognitive abilities to develop in puppies—just as it was one of the first social abilities to appear in us. This puppy power means that the day your eight-week-old puppy arrives, they already have the basic mind-reading abilities of a nine-month-old baby. The first time you point to a toy they have lost, or a treat they have missed, they can probably understand that you are trying to help them.[16]

Our comparison of wolves to dogs helped us see how the two species differ from each other, but it also pointed to important differences between individual puppies. The typical dog puppy performed better than the typical wolf puppy. But when we compared the dog puppies to one another, we saw that some made more correct choices than others.

A few dog puppies performed just as badly on these tests as the wolf puppies, while some were almost perfect. For example, on the marker test, Ashton and even cranky Dune got perfect scores with the new gestures, but Aiden only got 50 percent. Like a wolf puppy, Aiden was simply guessing. So just like we can see differences in the emotional reactions of individual dogs, we can also see individual differences in their cognition. These types of individual differences might help predict whether a puppy would make a successful service dog.

To find out we needed to test puppies, like Ashton, Aiden, and Dune, on a range of cognitive problems, and then evaluate whether their test performance was related to the type of great dog they would grow up to be.

CHAPTER 3

Celebrating Individuality

Class of Fall '19, left to right: Aries, Anya, Ying, Zina, Zax, Yolanda, and Weston

The class of Fall '19 has seven puppies. We have more than doubled our sample size, but the increase in complexity seems exponential. Seven little beds, seven bowls, seven poop-bag canisters—it is a lot to keep track of. Adding to the excitement, the puppies of Fall '19 all look the same. Unlike the distinguishing auburn locks of Ashton, the snowflake fur of Aiden, the dark coat of Dune—every puppy in the class of Fall '19 looks like they were copied and pasted from the pale puppy next to them. Which is almost true, at least for some of them, as we have three sets of siblings that might as well be twins. Puppies from the same litter are named with the same first letter. So the siblings are Aries and Anya, Ying and Yolanda, and Zax and Zina. Only Weston has no siblings, but he seems to be a patchwork of all of them; so, when he is running (and this class is always running) he could be anyone.

Fall '19 may look the same on the outside, but we know that each puppy will have important cognitive differences. Our goal is to find a way to measure these differences and see if individual cognitive profiles help us predict service dog success.

Before we started the Puppy Kindergarten, dog researchers primarily focused on temperament traits, testing how easily dogs startled, or how they responded to new people and places.[1] The effect of cognition on individuality is often vastly underestimated. This is likely due to an adherence to general intelligence theory, which states intelligence is something you have more or less of, like coffee in a cup. Also referred to as "g," "IQ," or "learning ability," general intelligence means that if someone is good at solving one type of problem, they are probably good at solving all types of problems.[2] So, if someone is good at memory games, they will probably be good at other games, even those that do not require memory. If this theory is correct, we can rank dogs on a linear scale, like the fastest runner or swimmer in the Olympics. According to general intelligence theory, cognition is a variable to consider but it's probably not doing much to create the individuality we see across dogs.

An alternative is the multiple intelligence theory, which states there are different kinds of cognitive abilities that vary across individuals. So just like a dog's hair color doesn't control what eye color they have, if a dog has a good sense of navigation, they are not necessarily good at reading gestures. A dog with good memory may not have good self-control. Each cognitive ability becomes like a letter in an alphabet. Just as letters can be combined and recombined to form multitudes of words, different skill levels across a variety of cognitive abilities can be combined and recombined to form multitudes of individuals.

Cognition at Work

Together with anthropologist Evan MacLean, our team at Duke tested over five hundred adult dogs, including military, service, and pet dogs, with twenty-five different cognitive games designed to test a wide variety of social and nonsocial abilities.[3]

If dogs have general intelligence like "IQ" or "learning ability" then performance in all the different games should be equally related to one another. This is not what we found. Instead, only performance in games that tested similar cognitive skills tended to cluster together. For example, when a dog performed well on one type of memory game, they tended to perform well on other memory games; however, good memory did not mean they had good self-control or vice versa. We identified a variety of cognitive abilities in

Original battery of 25 cognitive games

- Affect discrimination
- Arm pointing
- Visual discrimination
- Cylinder
- Detour navigation
- Spatial perseveration
- Social referencing
- Gaze direction
- Impossible task
- Working memory
- Sensory bias
- Marker gesture
- Odor discrimination

- Spatial transposition
- Distraction
- Contagious yawning
- Reaching
- Inferential reasoning
- Rotation
- Retrieval
- Laterality: first step
- Laterality: object
- Perspective taking
- Causal reasoning
- Transparent obstacle

The twenty-five cognitive games we played with 552 pet, service, and military dogs to test if dogs have one type of intelligence or many different types of cognitive abilities. Each circle represents one of the games.

dogs, including communicative requests, self-control, communicative comprehension, memory, object manipulation, and discrimination. We also found that performance on specific games was strongly associated with service and military dog success, including the ability to make eye contact, make inferences, follow a human pointing gesture, and exert self-control.* We were even able to use cognitive performance to make predictions about which of the dogs were most likely to successfully complete their training.[4]

Cognition is clearly a major player in generating individu-

---

* We are not the only research group to see this pattern. Animal behaviorist Lucia Lazarowski, who works with detection dogs at Auburn University, also found that cognitive abilities, motivation, and emotional responses are associated with training success in adult dogs. She found that lower trainer engagement, ignoring a misleading pointing gesture, weak self-control, and low levels of eye contact were most strongly associated with training success for detecting explosives.[5]

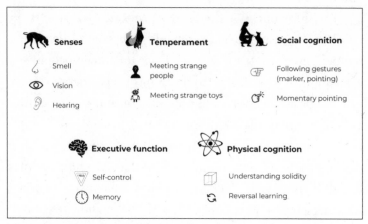

The Dog Cognitive Longitudinal Battery (DCLB) was designed to measure how dog puppy cognition matures during the end of rapid brain growth in dogs between eight and eighteen weeks of age. It measures the senses, temperament, social cognition, executive function, and physical cognition.

ality in our dogs and shapes how well they can solve a variety of real-world problems—including training for the complex tasks required to help the disabled or find explosives.

We'd made a good start, but to have the greatest impact on the supply of service dogs, we needed to make reliable predictions about puppies. So we reduced everything from our work with adult dogs down to puppy size. The space was smaller, testing sessions were shorter, breaks were longer. There were naps, before, during, and after the games. We turned the classroom into a magical wonderland, everything shrunk down to miniature proportions, and behind every door, a candy shop of treats and more toys than Santa's workshop. We created puppy versions of all the tests that had predicted training success in adults across different domains of cognition.

The first domain is the puppies' senses; we test to make sure they can see, hear, and smell. The second domain is temperament; we test how they react to strange people and even stranger toys. The third domain is social cognition—we test the basic theory-of-mind abilities related to cooperative communication. We include games with several levels of difficulty, from the marker test, which most puppies seem to pass easily, to a brief pointing gesture, which most puppies find difficult. We also include a game of fetch, because bringing back a toy for joint play indicates a willingness to cooperate. The fourth cognitive domain covers the abilities known as executive function. These are cognitive skills recruited to solve almost every type of problem. Our tests include measures of memory and self-control. The final cognitive domain is what we call "physics" because it tests how puppies understand the properties of the physical world. Our task measures if puppies can infer where a treat is hidden with knowledge that solid objects cannot pass through one another.

Each puppy will play the entire set of games every two weeks so we can chart any change in the puppies' performance during these ten weeks of rapid brain development. Like a powerful microscope, this will give us a high-resolution view into which cognitive skills develop at which age and in which order.

Before we start playing any of these cognitive games with the puppies, we quickly "rainbow" the kindergarten by assigning a different color to each puppy, including their collars, leashes, baskets, whiteboard markers, and folders for vet records.

But although they look the same, these puppies have

very distinct personalities. The first puppy who makes their presence known is Aries. Never has a puppy been named with more irony. The original Aries was a courageous winged ram in Greek mythology who rescued a young prince and was immortalized in a constellation. Aries the puppy is an absolute baby. For his first three weeks, Aries insists on being in someone's lap whenever possible. He has little interest in the other puppies. He wants to be with people. All the time. If the volunteers try to ignore him, Aries sits beside them, staring into their peripheral vision and making pathetic noises that we can only describe as a whiney-barky cry. It's impossible for the volunteers to resist.

His littermate, Anya, is the polar opposite. Although the smallest puppy of the group, she is fearless. She does not need to be petted and fussed over like her brother. She prefers to be obeyed. All day, she waits for the volunteers to bring her different toys, clean up her mess, and serve her meals. For the first week, she does not even eat in her kennel like the other puppies. She prefers to be hand-fed, one piece of kibble at a time, over the course of thirty minutes.

After a few weeks, there is another puppy we can spot a mile away. Ying is a good standard deviation larger than everyone else, and most of his size is in his belly. From week to week, we watch in astonishment as he balloons in height and weight. When he opens his mouth over his bowl, the food is sucked upward as if by a powerful vacuum.

Besides the gravitational force generated by his size, Ying has another superpower—his guilty-sad look.

Previously known as the guilty look, researchers have found that this expression does not mean dogs feel guilt.[6] Instead, it is a response to their owner's body language and tone of voice. The guilty expression comes from dogs' ability to raise a special inner muscle of their eyebrows. This muscle in dogs is considerably larger than it is in wolves, who never look guilty about anything.[7] The larger muscle allows dogs to raise their eyebrows and reveal more sclera, or the white part of their eyes, making them look simultaneously guilty and sad, which, in turn, seems to elicit a disproportionately positive response from humans.[8]

Many puppies perform the guilty-sad look with reasonable aplomb, but for Ying, it is an art form. His eyebrow muscles are formidable. No puppy has ever looked as pathetic or as devastated. Together with the fact that his size makes him clumsy—he is always tripping over his feet—Ying is the puppy who most often must "go see Margaret."

We call Margaret "the puppy vet," but this is a gross understatement of her talents. Margaret is a professor of veterinary medicine and behavior. She is one of only fifty registered veterinary behaviorists in the country, equivalent to a veterinary seventh dan black belt. She has a veterinary degree, a PhD, and a wide variety of research interests, including how to improve the lives of dogs, how dog cognition interacts with their sense of smell, and how vets can give better care. Margaret can do a blood draw between two licks of peanut butter. She can tell a urinary tract infection from the way a puppy squats, dermatitis from the way a puppy scratches. She prescribes fish oil when their skin is dry and probiotics when they have tummy troubles.

There would be no Puppy Kindergarten without Margaret. The co-director of our project, Margaret got us set up and started. But the puppies do not know how busy and important she is, or that other dogs come from all over the country to see her. She carefully examines every bump and scrape with a soft voice and gentle hands.

Ying especially loves to go see Margaret. This visit he comes into the playroom of the kindergarten, with its lavender smell and the sound of birds singing. Margaret joins him on the floor and asks him what's wrong. Ying unleashes his guilty-sad look and presents Margaret with the bump on his head that he got from walking into a door.

To her credit, Margaret never laughs. She runs her fingers over the bump, looks around the hair, applies pressure in different places, and tells Ying that she will keep an eye on his bump, but for now, he can go and play with the other puppies outside in the puppy park.

Bottom of the Class

Ying loves school. The classroom is right next to the front door, so before every bathroom break, he pulls hopefully toward class. He does the same at the end of recess or coming back from a walk. On the days when it is *finally* his turn, he bolts into the room so fast that he trips down the stairs and crashes into the puppy gate.

Kara smiles and lets him in. "Good morning, Ying."

Kara, our research coordinator, records and manages the data, and schedules the intricate ballet of puppies coming in and out of the classroom at just the right moment of their brain development. To the puppies, Kara is the magician on the stage. She is the one who plays all the games with them. She brings out the cylinder, the robots, and the other contraptions that mysteriously make treats appear.

"Are you ready, Ying?" Kara puts out the two bowls. Then she holds a treat toward Ying and says, "Puppy, look!"

Ying watches her every move. He sits up straight, a rapt expression on his face.

This is just a warm-up, and it's the easiest part of the games. It's simply to confirm that the puppies understand the game. All they have to do is watch Kara drop the treat in the bowl and go and get it.

"Okay!" Kara says after she has dropped the treat.

Without hesitating, Ying goes to the wrong bowl.

Kara sighs. "Poor Ying, you try so hard."

Despite loving school, Ying is not very good at it. When he is four months old, he fails the marker test, which other puppies can pass by eight weeks old. Some puppies are oblivious to whether they pass or fail. But Ying feels every missed treat like a stab to the heart and tries to climb into Kara's lap as though he were still a wee baby.

It is even more unfortunate for Ying because his sister Yolanda is gifted. She sails through the warm-ups and comes out 100 percent on the marker test. Yolanda is everything Ying is not. While he is huge and clumsy, she is elegant and beautiful. She also has surprisingly sharp teeth, and she is not afraid to use them.

It is a particular curse to have a sibling who succeeds so effortlessly. We give Ying extra cuddles and reassurance. But it means our approach is promising. Because at first, we

Ying, failing the marker test. Note the marker next to the bowl with the treat in it.

Yolanda and her shark toy

were worried that the games would be either too easy or too hard and all the dogs would pass or fail as both puppies and adults, which would not tell us much about the individuality we needed to measure. We had reason to be optimistic.

Before we opened the Puppy Kindergarten, our research team, led by psychologist Emily Bray,[*] tested 160 puppies when they were nine weeks old with the games we wanted to use in the kindergarten, then tested these same puppies again at two years old, using the adult dog version of the games.[†]

Emily found that in almost every game, dogs performed better as adults than as puppies. This news should be comforting to Ying, since it means even if he has a slow start, he will get better with time. The puppies had good memories, showed a measure of self-control, made some eye contact, and followed human gestures, but performed even better at these tasks as adults. When it came to self-control—the importance of which we will explore in our next chapter—the adult dogs made twice as many correct choices as they had when they were puppies. The same dogs made eye contact for three times longer as adults when faced with an unsolvable task or encouraged by a human. When it came to understanding human gestures, the adult dogs improved by more than 10 percent from their initial puppy performance.

---

[*] Emily started with us as a college student and is now a professor who has probably studied more service and guide dog puppies than anyone.
[†] The puppies came from sixty-five different litters and were raised in family homes by volunteer puppy raisers before they returned to Canine Companions for their professional training as adults. We did not raise or test these dogs as puppies or as adults at Duke. Instead, we worked at the Canine Companions headquarters in Santa Rosa.

These findings meant our games were doing what they were supposed to do: measuring a puppy's cognitive development.

Critically, Emily also found that a number of our games capture a snapshot of each puppy's individuality that anticipates what kind of dog they will grow up to be. Even though Ying's performance might improve, in certain games a puppy's adult performance was associated with what they had done in the past. A puppy like Yolanda, who scored relatively high in one of these games compared to Ying, will probably still score higher than Ying when they are both adults. The cognitive performance we observe in puppies on a range of measures remains relatively stable over time.[9] Given that we want to use a puppy's performance to predict their future success as service dogs, this suggests we're on the right track.

After only a short time, you will start to notice the kinds of traits that make your puppy special. Maybe they have an incredible guilty-sad look, like Ying, or maybe they think they are bigger than they are, like Anya, or maybe they whiney-barky cry unless you hold them all the time, like Aries. But it is not just these traits that make them special, it is also their distinct cognitive abilities. Ying's troubles with the marker test may mean he is not yet as skilled with human gestures as his sister, but he is by no means unintelligent. Instead, his success making eye contact with his adorable guilty looks suggests his cognitive strengths may be related to eliciting human attention and sympathy.

Similarly, your puppy may struggle to follow your gestures or forget you asked them to sit, but this does not mean they don't have more subtle cognitive strengths they rely on. What our dogs show us again and again is there are different ways they can demonstrate their intelligence even if they are not always brilliant in every way. It's what makes them so lovable. In the coming chapters, we will explore self-control, understanding of the physical world, and memory in our puppies to better understand how these abilities develop and create all the individuality that makes each dog unique.

CHAPTER 4

Control Yourself

Everything we ask of a puppy requires self-control: holding their bladder until they are in an appropriate outdoor space, resisting the urge to go very fast down a tall flight of stairs, or abstaining from jumping on furniture, counter surfing, or lunging across the road to chase a cat. To some extent we can try to exert control over puppy behavior using leashes and baby gates, but it's important to know when self-control appears and develops. You might be able to restrain your puppy with a leash when they are ten pounds, but this becomes more difficult when they are forty- to hundred-pound adults.

Congo, our puppy principal, demonstrates exceptional self-control.

When the class of Fall '19 arrives, Congo gets to work right away. He encourages the puppies to bite toys and not one another. He sets a calm, easy pace for walks. He never barks in reprimand. He demonstrates graceful play bows.

He sits patiently for treats and waits until every puppy bottom is firmly on the ground—the signal that treats may be dispensed. Seven puppies are a lot, even for him, and Congo comes home exhausted. But at work, he remains gentle, measured, and friendly.

For humans, self-control has many names. It is known as composure, discipline, and restraint. More generally, self-control is the look before the leap. The pause before you speak. Regardless of what we call it, self-control is a good example of a cognitive skill that is underrated but has an outsized effect on our lives. For example, one test of self-control is the marshmallow test, where researchers give a child a marshmallow and tell them they can either eat the single marshmallow right away or if they wait until the researcher returns, they can have many marshmallows. Some children eat the marshmallow immediately while others refuse to eat the single marshmallow for ten or even fifteen minutes while waiting for the researcher to return with a larger reward.[1, 2]

Studies have found that children who eat the marshmallow right away are more likely to struggle in school, have trouble paying attention, and have difficulty maintaining friends. Various follow-up studies found that when these same children grew up, they were more likely to be overweight, earn less money, and even have criminal records.[3]

Given everything we know about dogs, we suspected a dog's early ability to display self-control would have an impact in the outcome of service dog success. But how would you measure self-control in puppies? One answer might be to see how fast puppies could be trained, since puppy training often requires self-control. However, training relies on a host of other cognitive skills too—like memory and communicative abilities.

Instead, we invented a kind of marshmallow test for animals that we can use with puppies. While the puppy watches, we put a treat inside a clear plastic cylinder, open at both ends. As a warm-up, we cover the cylinder in a cloth, so it is opaque. This way the puppies are introduced to obtaining the food after it disappears into the cylinder. They simply go around to one of the open ends to retrieve it. Next is the test. We remove the cloth from the cylinder, so that it becomes transparent, and they can see the treat inside.

You might think giving the puppies more information by letting them see the treat inside the cylinder would make the task easier, but it actually creates a cognitive conflict. They remember they had to make a detour around the cylinder and get the treat from one of the open ends, but because the cylinder is now transparent it *looks* like they could just directly approach and grab it. In order to win the game, each puppy must avoid the temptation of approaching and bumping into the cylinder, and instead go around to the open end. To succeed, they must resist what their eyes are telling them and continue making the detour they remember making previously. This takes self-control.

Anya bumps into the cylinder a few times, but on the final trial, her self-control wins out and she manages to take the detour around the side and get the treat. Aries, our little Greek ram, has nearly perfect self-restraint. He dances gracefully toward the cylinder, then almost timidly reaches around and delicately extracts the treat with his lips. Ying bumps straight into the cylinder. Weston also fails spectacularly. Weston is our most confident puppy. He enters every public space—the puppy park, the classroom, the hallway—as though he is entering a stadium full of cheering fans. When Weston sees the cylinder, he gets so excited to retrieve the treat that he dislodges the whole apparatus during the warm-ups.

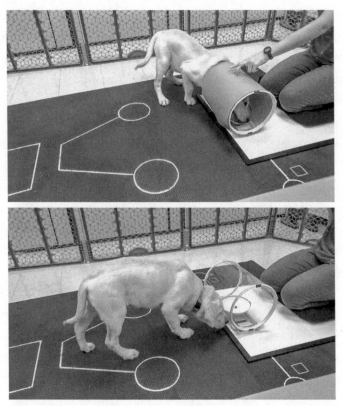

The first photo is Ying during the warm-ups, where the cylinder is covered with a plastic cloth. The second photo is Ying bumping into the cylinder once the cover is removed.

To be fair to Ying and Weston, other animals struggled with this test. We recruited researchers from all over the world to use our cylinder test with over five hundred and fifty animals from thirty-six different species, including birds, apes, monkeys, dogs, lemurs, and elephants.[4]

Most species went directly for the food and bumped into the solid cylinder even though they knew from their warm-ups that they could only retrieve the food through the open ends. A few species, like the great apes, had the self-control to inhibit this reaching response and made no mistakes. Species like squirrel monkeys, on the other hand, never learned—even after ten chances. It turns out that the animals who pass our marshmallow test have brains with more raw computing power. Smaller brained animals struggle with self-control while larger brained animals, like great apes, master the test almost immediately. Ying's performance is closer to that of a squirrel monkey than a great ape.

But then, when he turns sixteen weeks old, Ying transforms. Kara baits the cylinder, removes the cover, and . . . voilà! On the first trial, Ying goes directly around the side and gets the treat—without even touching the front of the cylinder.

It is not just a fluke. Now Ying does it every time. It is a delight to watch him delicately detour around and then thump his tail in response to all the hugs and praise. At sixteen weeks the executive center in Ying's brain manages to override his impulse to approach directly. He can go around the cylinder to get the treat. At eighteen weeks, he passes a harder version of the test where we block the end he prefers

to use, so that he has to show even *more* restraint. It looks like, at least with Ying, self-control is something that develops over time.

Weston, on the other hand, has no such luck. During the warm-up when the cylinder is still opaque, he just bounces right up to it and punches it with his paw.

An uncomfortable pause ensues. No one fails so quickly. In our giant species comparison, even *pigeons* pass this part. Weston keeps punching the cylinder with his paw. He can't help himself.

When we take off the covering, it gets worse. Weston does not bump into the cylinder with his nose. He charges it. He jumps on top of it. He rips it off the base and it rolls across the floor.

In the weeks that follow, Weston continuously manages to dislodge the cylinder. As he grows older and larger, instead of improving, like Ying, he just gets better at ripping the cylinder off its base and throwing it across the room.

Shark Week

On average, we have found that puppies do not start passing the self-control test until they are around ten weeks old, and do not achieve mastery until they are fourteen weeks old. Which means that when you first bring your eight-week-old puppy home, they have relatively little self-control. At the Puppy Kindergarten, we don't do much with puppies at this age other than try to teach them their names and how to sit. During this eight-to-ten-week period when puppies lack self-control abilities, we also see the most indoor accidents occur. Instead of expecting the puppies to know when they start feeling like they might want to go to the bathroom and then holding it until they are outside, we try to anticipate when they need to go to the bathroom by taking them out every thirty minutes, plus when they wake up in the morning and after naps, after they drink water, and twenty minutes after they eat. Puppies do not usually pee while you hold them, so a neat trick is to carry your puppy to the bathroom once they wake up from a nap.

However, even in the class of Fall '19, we can see that some puppies have perfect self-control, others have self-control that gradually improves over time, and some puppies utterly lack the ability to control themselves in any scenario.

In the kindergarten, the importance of self-control becomes very clear when the puppies are twelve weeks old and they all start teething at the same time. At around three weeks old, puppies get their milk teeth. At around twelve weeks old, puppies start losing their milk teeth and their permanent teeth come in. Usually, puppies just swallow their milk teeth as they fall out, but with seven puppies teething at once, the volunteers invariably find tiny teeth on the floor, like shark teeth on a beach.

It can take six months for the full set of permanent teeth to come in, which is why puppies are famous for chewing on anything and everything, including one another. We use a range of teething toys, from synthetic bones that don't break off into choking hazards to frozen chew toys that soothe their gums. The puppies tend to like it if you hold one end of the chew toy in your fist so they can really grind their back molars—sometimes this puts them into a kind of rhythmic chewing trance.

Unfortunately, the beginning of the permanent teeth at twelve weeks coincides with an uptick in puppy energy. The puppies go from droopy eight-week-olds who are likely to fall asleep on their feet, to twelve-week-olds powered by rocket fuel. The intensive teething, increase in energy, and increasing amounts of self-control are why we tend to start training in earnest at around this twelve-week mark, in particular, teaching them manners. Manners are all about overriding first impulses. Like stuffing food into your mouth. Or taking something you want without asking. A polite "sit" requires self-control because the puppies must stop doing what they want, which includes jumping up on people, trying to tip over the treat bin, or playing too roughly on the playground.

We only use positive reinforcement with the puppies. No puppy is shouted at or punished. Positive reinforcement is not only about rewarding the behavior you hope to see repeated, but it also gets a puppy to pit one desire against the other and rely on self-control to make the correct choice. If a puppy is really enjoying tearing into a section of the wall, it doesn't make sense to offer them something that is less appealing—like a sip of water. We use kibble to teach the puppies manners, to sit and not jump, to refrain from chewing people's trouser legs, and to resist pulling their volunteer behind them when they see Congo or a puppy classmate. If a puppy is doing something naughty, we use kibble to distract them. We use kibble to coax the puppies to go to the bathroom in the pouring rain. If kibble is not enough, we use small, low-calorie treats, like Zuke's. If that doesn't work, we move on to small soft treats, like Little-Jacs, and if that doesn't work, we move on to the very top of the treat chain— Cheez Whiz. Cheez Whiz is the ultimate enticement, and it will nearly always stop a puppy in its tracks and give them the chance to make a different choice.

Each puppy receives a training sheet with a list of skills they must master.

Skill	Proficiency (I, P, or N)*
Name (eye contact)	8 weeks
Sit	10 weeks
Hurry	Start at 8 weeks, and expect success by 16 weeks
Kennel	10–11 weeks
Down	12 weeks

Skill	Proficiency (I, P, or N)*
Let's go	16 weeks
Here	14 weeks
Wait	16 weeks
Heel	18 weeks
Leash lure	16 weeks
Cape	14 weeks
Doggy leader	14 weeks
Nail clip sensitivity	Start habituating at 10 weeks, and expect success by 18 weeks
Ear cleaning	Start habituating at 10 weeks, and expect success by 18 weeks
Toothbrushing	Start habituating at 10 weeks, and expect success by 18 weeks

*I = Introduced/Learning, P = Proficient, N = Not Introduced

Midterms for the puppies fall around their fourteen-week-old mark—which is when most puppies are able to pass the more difficult version of the self-control test—and allow us to see where they are in terms of training and understanding skills. At the end of the semester, when the puppies are eighteen weeks old and about to graduate from kindergarten, they will have their final exam. The goal for each puppy is to be proficient in all skills.

Self-control is not just about training. It is also about inhibiting behavior. The puppy exhibiting the worst behavior in the class of Fall '19 also happens to be the puppy with the least self-control: Weston. By the time the puppies are twelve weeks old and teething, Weston starts every day with an insane session of zoomies in the puppy park. He kicks sleepy puppies in the face, nips their bottoms—anything to prevent them from being boring. Ying usually hides under a bench and pretends he is not there, but his sister Yolanda will meet Weston head-on. Anya, even though she is the youngest and the smallest, invariably joins in and soon the three of them are riled up, tackling one another, growling, and biting, while the poor volunteers try to keep everything under control. There are basic rules for the playground (no biting, no fierce growling, no bullying) and Weston breaks all of them. Only fifteen minutes into recess and the volunteers are exhausted.

Weston's teeth marks show up in the drywall of the kindergarten, on the nap-time crates, and in the ankles of the volunteers' jeans. One of his baby teeth actually gets lodged in a volunteer's sweater. Weston just grins a bloody grin and flips over a water bowl.

"The only time he is good is when he's asleep," says a volunteer with dark circles under her eyes. Which gives us an idea.

The Long Walk

When a puppy has little or no self-control, the only tool we have left is to induce fatigue. But how do you tire out a puppy whose utter lack of self-restraint means he plays too rough with other puppies, runs around like a maniac, and chases everything that moves? Plus, we need a puppy like Weston to be able to walk nicely on a loose leash. This would be difficult even for a puppy with incredible self-control. Luckily, Canine Companions has the doggy leader, a magic training tool that works under the assumption that self-control develops at different speeds in different puppies at different times.

The doggy leader loops over a puppy's nose and functions like the bridle on a horse so that puppies can only walk in the direction their head is facing. This means that you can gently guide the puppy's movement by controlling the direction their head is facing. If a puppy tugs while wearing a doggy leader, their nose is pulled back toward you and they stop.

While we would like to tell you that you just slip the doggy leader on with instantaneous results, the reality is that the hard work for the doggy leader is front loaded. It takes weeks to train a puppy to wear it properly, using lots of praise and treats and patience.

The first time Weston puts on the doggy leader, he is outraged.

He rubs his nose on the ground, pouts, and sulks. He pancakes on the floor and refuses to move. It takes several weeks, tons of treats, and endless patience before Weston starts walking with the doggy leader.

The outcome of those weeks of coaxing and encouragement is that by fourteen weeks old, Weston is walking on a leash like a pro. No pulling, no tugging, and no nose rubbing. If Weston starts zoomies at recess, he goes out for a walk. If he wakes up after his nap and starts harassing the other puppies, he goes out for a walk. Weston walks on hot September afternoons and freezing mornings in November. When it is pouring rain, he walks through the long winding corridors of the biological building's sub-basement. After a solid walk, Weston arrives back at the Puppy Kindergarten, if not exactly exhausted, then more subdued. He might even nap. By the time he is sixteen weeks old, he is walking several miles a day.

One more word about Weston's personality that will help explain his results in the self-control games. On a sunny afternoon, we take the puppies to meet the Duke basketball team. These humans are, without question, the tallest people the puppies have ever seen. From a puppy's point of view, they must seem like giants. All the puppies roll over, submissively exposing their bellies. While on their backs, they wag their tails so hard, they propel themselves into the players' massive feet, and they are scooped into muscled arms and cooed over like babies.

Not Weston. He struts straight past the players to the giant Duke *D* on the court. And pees on it.

Weston is the middle puppy, with his bottom hanging down.

These Dogs Are the Bomb

When we talk about cognitive skills, we try not to talk about them in terms of passing or failing. There is a cognitive profile that is a better fit for a life in service, but that does not mean that a different cognitive profile is somehow wrong. Weston may seem like a failure, but for some working dogs, failure in the self-control games predicts success.

We tested hundreds of military bomb detection dogs.[5] These military dogs are trained to perform a highly specialized job, just like service dogs. They're also, like the Canine Companions dogs, primarily Labrador retrievers. But the similarities end there. Canine Companions dogs have names like Rainbow, Aurora, and Wisdom. The military dogs have names like Storm, Tank, and Captain. A Canine Companions dog can snooze for most of the day. The military dogs are live wires, crackling with energy, full of what their train-ers call "drive."

Military bomb detection dogs work by entering an area with their handler. The dog searches the area first. When the dog locates the target substance, they "alert" the handler by sitting down or staring at the target—without barking and, most important, without touching the target.[6]

However, the scent work of military dogs is only a small part of the cognitive skills they need to be successful. For instance, if there is a ravine or a pothole, the dog has to mentally represent space and take a detour around the ob-stacle. They have to sort through odors that might mask the target odor. They have to remember where they searched before and avoid distractions like civilians and free-ranging dogs. They must retain the memory of different categories of explosives (TNT, potassium chlorate).

When we tested these dogs, we quickly discovered an-other trait that led to detection dog success—they need to have near zero self-control. While service dogs prefer treats, the military dogs *love* toys. So instead of hiding food during our games, we hid small squeaky yellow ducks.

Unlike all the other Labradors we had tested, the mili-tary dogs did not play with the ducks—they just swallowed them whole. Shocked, we stopped testing and alerted the veterinarian staff. The dogs passed the ducks through their digestive systems a couple of days later without a problem, but we had learned our lesson. We replaced the ducks with larger rubber toys.

It was not just the ducks that were in danger. Late one night, our students called with an unusual request. They wanted to know if they could get some bike helmets for the detour game. During this game, the dog had to take a detour around a small fence to get a reward. Instead of taking the detour, the military dogs were barreling straight through the fence. Our students were worried someone would get hurt when the fence and the dog came crashing down on the person who was running the game.

But this is what makes military dogs successful. They are like soldiers; when everyone else is running away from danger, they run toward it. On the battlefield, second thoughts are a liability. Military dogs have taught us that getting the answer wrong in a game does not mean the dog has failed. It just means they have a different way of solving problems.

In the meantime, all the long walks and nap times pay off, and Weston develops a little more self-control as he gets older. But Weston still does not seem suited to be a service dog. It's okay. Weston helps illustrate how cognitive profiles are useful since they give us clues about the type of job a dog might do well. Weston, who needs an active lifestyle, goes on to live with journalist and local Raleigh celebrity Monica Laliberte. Monica has raised at least six Canine Companions puppies, and Weston is her new project. Like Weston, Monica is fast and fearless, and she recognizes he will be a fantastic news desk dog. Right on the front lines, in on all the action. It is the perfect job for him.

CHAPTER 5

A Surprising Marker of Success

The self-control games taught us that the way each puppy solves problems can be an asset in some situations and a liability in others. How can we know whether a dog's cognitive profile will help or hinder them? Part of the challenge is recognizing how our own preconceptions can distort these kinds of evaluations.

We've heard people say, "Oh, my dog could never be a service dog, he's too lazy." But the ability to lie calmly in a doctor's office or school for hours is critical to service dog success. We've also heard, "My dog is hyper, he could never do anything useful." Yet one of the key criteria for a military dog is unrelenting drive—they sure look hyper when they are in the field, but their high energy level has been shaped into a sophisticated talent. In the class of Fall '19, no one illustrates this better than Zax.

Right from the start, Zax is as cool as a cucumber. Un-

Zax, helping Margaret train medical students on how to give pediatric exams. A child on all fours has their organs in roughly the same place as a puppy, so Zax teaches them how to examine small cute creatures who may wiggle.

like Weston, who takes up all the space in a room, Zax is content to slip under the radar, just another pale face in the blur of pale faces. And unlike Ying, who has the body of a lion cub and the heart of a coward, Zax has the quiet confidence of a natural leader. Handsome and charming, Zax is the one who gently greets small children who seem nervous. He is the one who calms grown-ups who might have had a bad experience with dogs. Zax cannot be provoked. Other puppies crash into him, take his toys, splash water in his face—he just gets up and takes himself somewhere else.

In the cognitive games, it is clear Zax is gifted. On the marker test he is 100 percent every time. Even on the harder version, he is the top of the class, every week. He has the self-control of a Jedi master. He checks every box of what you might think a service dog should look and behave like— with one major exception.

He eats his own poop.

A flaw that obviously portends total failure.

You can tell when it's happening because the sound of disgusted young people sails through the air. *Gross! Zax! Again?*

Zax will eat poop that has frozen overnight. He will eat poop warm out of another puppy's bottom. After he goes to the bathroom, he will swiftly spin around and eat his own. The volunteers are so distraught that they develop a buddy system. Every time Zax goes to the bathroom, one volunteer holds his leash while a second volunteer scoops the poop as soon as it hits the ground.

Poop eating, or coprophagy, is a concern we often hear about. It's usually whispered shamefully, out of other people's hearing. "How to stop my dog from eating poop" is one of the most googled dog-related questions on the internet. Zax is not allowed to eat poop. Besides the risk of infection, it is unsightly, and it makes his breath smell. But poop eating is a common behavior among mammals because it can be beneficial. Mother dogs eat the feces of their newborn puppies to keep their dens clean. Other canines, like coyotes, eat the poop of rival coyotes and replace it with their own to mark their territory. The poop of hooved animals, like deer and cows, is only partially digested (50–60 percent), and the antioxidants and other nutrients in their poop have been found to be beneficial to the immune system.[1] But since coprophagy risks infection from parasites or harmful pathogens, eating poop of any kind, even herbivore poop, is obviously not recommended.

There is not much research on how to stop coprophagy;

the one study we found ranked various preventative measures, such as sprinkling the poop with hot sauce or closing your eyes and hoping the behavior goes away. The study showed that commercial deterrents work just marginally better than doing nothing.[2] The only way we could stop Zax's poop eating, or at least reduce his intake level, was to take preventative measures: Keep him on a leash, scoop the poop as soon as it happens, and stay vigilant.

Though all the volunteers are convinced that Zax's poop eating absolutely disqualifies him from ever graduating as a service dog, it might actually be a sign that he will. When Emily Bray analyzed the data of surveys filled out by 138 guide dog puppy raisers, she found that dogs who eat poop are actually *more* likely to graduate as service dogs.[*] And it isn't just a slight correlation. Coprophagy is one of the top five behaviors that predict a Canine Companions dog is likely to graduate.[3] One hypothesis is that being a Labrador retriever, Zax is susceptible to a mutation in the pro-opiomelanocortin (POMC) gene. This mutation disrupts the coding sequence of beta-endorphins and a neuropeptide called beta-MSH, and creates an insatiable appetite.[4] POMC is the reason why Labs tend to become overweight (a mutation in the same gene is linked to obesity in people). It is also why Labs will eat the fuzz off socks, an entire bin of chicken food, a bag of wet cement—the list goes on. Geneticists have found that the mutation of this gene appears more frequently in dogs

---

[*] Emily analyzed C-BARQ surveys; these are temperament and behavior questionnaires that assess everything from energy levels to barking to tail chasing.

selected to become service dogs than in regular pet dogs. So, Zax's kryptonite may well be his strength. Zax eats poop because he's hungry enough to eat anything. This quality makes Zax super trainable. He will use all his superpowers for a piece of kibble.

Eating poop is an example of how our preconceptions can be wrong and what looks like a liability can actually be an asset.

Physics

Preconceptions work the other way too. There are cognitive skills that we might guess are crucial to service dog success but our research suggests they have the opposite effect. Take physics, for example. We tend to think of physics as being one of the "smart" sciences, like computer science or mathematics. Physics tests our understanding of the physical world. It can be as complicated as string theory or as simple as knowing that gravity causes objects to fall.

There are a million little physics tests we pass every day that involve understanding properties of matter, inertia, electrical energy, light refraction, weather patterns. By two years old, we have formed expectations about the way certain matter behaves. Some objects float in water while others sink. Someone cannot be in two places at the same time.

In general, dogs do not tend to excel at physics in the same way.

They bump into doors, get stuck in small spaces, try to pull a stick through vertical posts in a fence.[5] Dogs often fail to understand that objects fall, balloons pop, and waves

Puppies failing to understand the physical principal of connectivity. Zax (left) is the only one going in the right direction. Note Congo (far left) modeling infinite patience.

crash.* But just like some of us are born to be rocket scientists while others will bump our heads on the same closet door for decades, some dogs are simply better at understanding the physical properties of the world than others.

Zax loves physics. He watches Kara like she's his favorite TV show. He sits up straight, eyes locked on her every move as she tests whether he understands that solid objects hold their shape and do not pass through one another.

Kara drops a treat into a deep bowl that is attached to the top of a wooden block. Then the magic begins. Kara brings out her screen and hides the bowl. With a flourish, she lifts

---

* Gravity is a good example. You might have seen a dog drop a ball on a steep hill then act surprised when the ball rolls away. Dogs do seem to understand that objects fall, but this understanding is limited.[6]

a yellow cloth above the screen, letting it flutter so Zax can just see it, and then places it over the bowl behind the screen. Kara takes another identical yellow circle, again flourishes it just so and lets it float down onto the mat out of Zax's sight.

A puppy (not Zax) watches as Kara prepares the physics test.

Kara removes the screen to reveal one of the yellow circles is lying flat on the ground while the other is draped over something about the same shape and size as the bowl and block where the treat was hidden. If you understand that solid objects cannot pass through one another, and that they maintain their shapes even when you cannot see them, it is obvious where to search for the treat. If you have no concept of solidity, you will be flummoxed. As humans, it is hard to imagine failing this test, but most young puppies struggle with it. The day before, Weston failed. So did Yolanda. To their credit, many adult dogs also do not pass this on their first or even second try.

Zax pauses a moment. Then trots over to the bowl-shaped yellow cloth. He seems to get the whole solid-object thing. Einstein. You might think that doing well in physics is an asset. However, we have found being good at this particular game means, if anything, that a dog is *less* likely to succeed as a service dog.[7]

Puppies like Zax show us that predicting who will make it is complicated. It is not enough to trust preconceptions and intuitions. For example, every semester, the volunteers predict who will make it as a service dog. For Fall '19, there is a clear winner: Anya.

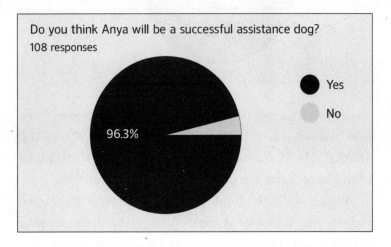

Anya, our Russian princess, is the one puppy everyone thinks will make it.

Our volunteers describe her as sweet and clever and full of heart:

She's very sociable but also a quick learner and knows her boundaries. She's a princess and a warrior at heart.

Angel child, she's so friendly and calm.

She can be sassy sometimes, but I think she has what it takes.

Then there are the puppies where expectations are not so high—like Anya's brother Aries, the Greek ram himself.

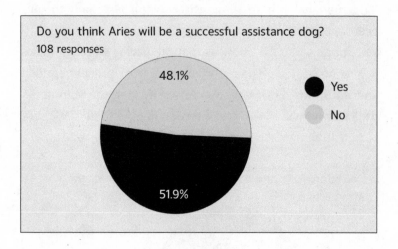

Do you think Aries will be a successful assistance dog?
108 responses

48.1%

51.9%

● Yes

No

By the time our puppies graduate the kindergarten, we want them to know basic skills, like SIT, DOWN, WAIT, SHAKE, HERE. Aries knows none of them. Or rather, he knows all of them, but doesn't want to do any of them.

We aim to have all our puppies perfectly toilet-trained and sleeping through the night. Aries passed toilet-training but barks in his crate. It's not that he is hungry or needs to go to the bathroom. He just imagines that he should be spread out in bed beside someone, his head on a pillow.

Aries will not walk with his doggy leader. This is bewildering to us because by the time they graduate, all our puppies walk on their doggy leader. Aries refuses. He sits down and does not budge. Usually, you can wait a puppy out. You can upgrade the treats. Not Aries. He wants to be picked up and carried. And that's that.

It is astounding how much the volunteers love him. For

the class of Fall '19, we launch the Favorite Puppy Competition. Preceded by much emotion and discussion, you would think the winner would be Weston, Mr. Confident, or Anya, the princess everyone thinks will succeed.

Instead, it is a dead tie between Aries and Ying. When the results are put together with our other competitions (Best Halloween Costume, Best School Portrait), Aries comes out ahead by a slight margin.

Aries on top of Yolanda on top of a lap

The volunteers love him despite how difficult he is, or perhaps because of it. And while no one can see him pulling a wheelchair, everyone can see him spending the rest of his life making someone happy. Something about Aries just makes him a great dog. Just like Ashton, or Zax, Aries is yet another example of the fact that there is no "right" or "wrong" cognitive profile. Each individual puppy interprets the world their own way and solves problems differently. When we make assumptions about who our puppies are destined to become, we are almost always wrong. Puppies who are good at the physics game—at understanding a complex concept that we as humans think of as "intelligence"—are usually less likely to succeed as service dogs. Conversely, puppies who eat poop, a behavior we think of as unseemly for a service dog, tend to grow up to be successful. At the Puppy Kindergarten, our goal is to help realize each puppy's potential, and so we investigate how to do just that. We focus on understanding each dog's cognitive profile, but we also want to know how we can help them succeed. Our next step is to investigate how a puppy's genes and environment interact to shape their behavior. We know all behavior is always shaped by both, but can we learn the degree to which each factor contributes to a dog's success?

CHAPTER 6

Brainy Genes

You don't get to choose members of your family, except when it comes to your dog. Half the fun of getting a puppy is imagining how they will grow into your life and get along with the people you love. Do you enjoy cross-country running or duck hunting? Do you live with children or elderly parents? Do you need a dog who can fit under an airplane seat or a dog who can pull a sled? Most likely you have a breed in mind. Either a breed of dog you had growing up, a breed of dog you have admired from a distance, or maybe one day you fell down the rabbit hole of Instagram and decided that your life is incomplete without a cockapoo.

When people visit the kindergarten, they always ask what breed our puppies are. When we reveal that our dogs are some mix of Labrador and golden retriever, they usually nod their heads. "Labs are great dogs," they might say. "Super friendly. Good with kids."

We often wonder what they would think if they could see the military dogs, who are also Labrador retrievers, but so temperamentally and cognitively distinct from service dogs that they might as well be a different species.[1] And as we saw with Aries and Anya, even puppies from the same litter can be very different.

When people discuss breeds, what they are really talking about is genetics—the physical, behavioral, and cognitive variation they believe is reliably passed down from one generation to the next through genes. Many morphological characteristics clearly differentiate between populations of dogs. All Great Danes are big and all Chihuahuas are small. Labradors have floppy ears while German shepherds have pointy ears. Dogs have an unusual number of stark morphological traits like those caused by very simple genetic differences.[2] In most species, including humans, the vast majority of traits are controlled by a network of hundreds or thousands of genes. But even traits controlled by relatively simple genetic differences, such as those that differentiate one breed from the next, show a degree of individual variability. For example, while all Great Danes are large, their weight can range from 110 to 175 pounds. That individual variation is due to the version of the height genes a dog inherits and the type of environment a dog lives in, which shapes how those height genes are expressed.[3] All individual variation in all types of traits, including cognitive and behavioral ones, is always a product of genes interacting with the environment. The challenge becomes assessing heritability, or the relative degree of influence genes and environment have on each other as they interact to shape an individual. When traits have high heritability, genetic differences between individuals explain a high proportion of their individual differences—and environment may matter less. In these cases, with further genetic work, a genetic mechanism can often be identified that has great influence on shaping a trait of interest. Traits showing a high degree of heritability also make promising targets for artificial selection—selection that can potentially increase the prevalence of a desirable trait within a population.[4]

This happens to be exactly what we are interested in at the Puppy Kindergarten—the amount of cognitive variation that can be attributed to genetic variation as opposed to environmental factors (e.g., rearing, training, etc.). Highly heritable cognitive traits might provide an additional avenue to increase the supply of service dogs.

Classic Lab

The puppies of Fall '19 graduate, go to their new puppy raisers, and by the time we have waved them off, cleaned out the kindergarten, ordered new supplies, and taken a breath—the holidays have passed, and it is time for the new class to arrive.

Meet the class of Spring '20. Arthur is the Labbiest Lab we have ever had. Labrador retrievers were the most popular breed of dog in the United States for over thirty years.[*] De-

---

[*] In 2023, they were surpassed by French bulldogs.

spite popular assumption, their ancestors were not from Labrador (the easternmost province of Canada) but from an island called Newfoundland—a short subarctic trip southeast. They were brought to England in the 1800s and reportedly bred as fishing dogs, helping fishermen retrieve fish or nets that fell into the water. The dogs were strong swimmers, with a dense, water-repellent coat that didn't freeze in the icy waters. By the time Labrador retrievers arrived in the United States, they were prized as gun dogs, especially for duck hunting. The dogs had "soft mouths," which meant they would not damage their quarry while retrieving it.

Due to the breed's popularity, most people who visit the kindergarten either have a Lab or have had one, and when it comes to the class of Spring '20, visitors gravitate toward Arthur. Arthur is the Lab of popular imagination. He is black, just like those original Newfoundland dogs, with a short, water-repellent coat that shines with natural oils. When we fill up the puppy pool on a warm day, some of the other puppies are wary of getting their feet wet. Arthur jumps in without hesitation, clearly a natural swimmer. He is plump without being chubby, silly without being clumsy, and his ears flop down to the perfect length just below his chin. He has a strong jaw, but a gentle mouth. While the other puppies leave teeth marks in everything from metal water bowls to the hands of the volunteers, Arthur can pick up his favorite teddy bear without messing up its fur.

Retrieval is one of the skills that predicts whether a dog will graduate as a service dog, so we always include a game of fetch when the puppies are in the classroom. We take notes on whether they chase the toy, touch it a little, or bring it back for us to throw again. Even though all our puppies are at least part Labrador retriever, some of them have no interest in retrieving at all. At eight weeks old, Arthur has the highest retrieval score of his class.

Every now and then, as Arthur stands tall and sniffs the air, or pads across the puppy park, you can see what a calm and dignified dog he will become. How handsome and gentle he will be. And then, just as you are admiring his regal grace, he will fall into a play bow and stick out his little pink tongue.

It kills the volunteers every time.

A Breed by Any Other Name

Until the twentieth century, dogs were bred for their jobs. Each breed's name corresponds to their purpose. Any dog that guarded sheep was a shepherd, any dog that chased hares was a harrier, and any lapdog was a spaniel. There were dozens of jobs that dogs were bred to perform, ranging from guards and guides to garbage disposers.[5] Over time, the job each breed was meant to perform gave them a certain appearance. Labradors, for instance, were characterized by their size—large and powerful enough to swim in the open ocean, but small enough to haul fish into a boat; a coat that was water resistant and dried fast; and the broad otter tail that helped them swim.[6]

Another example is the English bulldog. In the seventeenth century, butchers were required by law to tie a bull to a stake and unleash a group of dogs to kill it because people (mistakenly) believed this tenderized the meat. Reportedly, any dog brave enough to launch itself at an enraged bull was termed a "bulldog," but in general, it was helpful for the dogs to be low to the ground, since bulls charge with their heads down. The dogs had to have strong jaws so they could lock onto the tender nose of the bull and not have their teeth torn out when the bull flung them around. To breathe during all of this, it was advantageous to have a protruding underjaw and wide flared nostrils. Fortunately, this inhumane practice was outlawed in 1835. But to this day, bulldogs have retained the traits of bullbaiting—the underbite, low height, and powerful-looking jaw.[7]

Until the nineteenth century, there were only a few

dozen dog breeds, and no one cared much what the dogs looked like; it was more important they did the jobs they were bred for. The four hundred or so dog breeds we know today, also known as "modern breeds," were created by a seismic cultural shift that began with the industrial revolution. In addition to the rural population that flooded into cities to create the working class, the industrial revolution created a new middle class. These people worked office jobs, lived in the city, and looked to aristocratic celebrities for fashion and style. This new middle class didn't need their dogs to work. Instead, they wanted a dog who would make a good pet, but who could also function like a designer handbag, and advertise the taste and disposable income of the owner.[8]

Dog breeding quickly became a Victorian obsession. It must have been wonderful, and perhaps magical, to watch dogs grow hair to their shoulders, bushy eyebrows, or even a mustache. You could shrink a dog to the size of a kitten or grow one to the size of a pony. You could create a dog who stood like a prince, or a dog with the face of a goblin.

The first formal dog show was held on June 28, 1859. There were only two classes: pointers and setters. Four years later, the show exploded, with over a thousand entries.

Dog shows were the place where the nouveau riche came to throw their money around. While you could buy a sheepdog for a pound, a first-class show collie was worth up to a thousand pounds, roughly one hundred and twenty thousand dollars today.

As you can imagine, this kind of money attracted some unscrupulous people. With a little trimming and shoe pol-

Dog fashions for 1889. *Photo:* Punch *Magazine*

ish, a sheepdog might be passed off as a show collie, and by the time the scam was discovered, the perpetrators were long gone. In response to this, the first Kennel Club was set up in 1873 to establish the identity and descent of pedigree dogs.

And so, dog lovers blindly began one of the greatest genetic experiments in history.

Next Generation

In the mid-1800s, genes were not yet discovered. It was known that traits were passed down from generation to generation, but there was little understanding of how this worked. Darwin thought the body's cells emitted particles of inheritance called "gemmules." Gregor Mendel was closer to the truth when he wrote about his experiments with pea plants, but his writings of the 1850s were not discovered by the scientific community until half a century later. Thus DNA was not discovered and described until well into the 1900s.[9]

By then dog breeders and kennel clubs had created a standardized list of physical traits for each dog breed. Since every dog in a certain breed has the same list of physical traits, which can be reliably observed by anyone, geneticists could see where the appearance of one breed differed from another breed and, in a few rare and lucky cases, pinpoint the gene or genes responsible for this trait. For humans, hundreds of genes are implicated in determining height. In dogs, a single gene, IGF1, is a major factor in making small dogs small, and another gene, IGSF1, is a major factor making tall dogs tall.[10] You don't need to lay out the entire three billion or so base pairs of the genome of each dog breed to see where they differ. Instead, you can look at different snapshots of variation in the genome, called "single nucleotide polymorphisms," or SNPs. Using this method, geneticists have found that the mustache and eyebrows of the schnauzer are associated with a mutation in the gene called RSPO2.[11]

Because breeds have been so carefully selected for appearance, you can be reasonably sure that when you choose a breed, like a Labrador retriever, those dogs will have similar physical characteristics. All the puppies at the Puppy Kindergarten, and most other service dogs, are Labrador retrievers, or a Lab–Golden retriever mix. Because we know roughly how large they will be, we know they will be physically capable of doing the tasks they might have to do. For example, no miniature poodle, no matter how cognitively gifted, will be able to turn on a light switch—they are just too short. And no Chihuahua, no matter how brilliant, will be able to pull a wheelchair.

And even though any medium to large breed could be a service dog (German shepherds were the original guide dogs), Canine Companions and other service dog organizations choose Labrador retrievers because that's what the general public *thinks* a service dog should look like. It is already difficult enough for a disabled person to gain access to certain spaces. If they must also overcome people's preconceptions and explain that a Rottweiler can be a service dog, it just makes things needlessly difficult.

But when researching breeds as a potential puppy parent, keep in mind that the only characteristic that can be guaranteed when you go to a breeder is physical appearance. The breed tells you very little about who the puppy will grow up to be.

It may seem surprising that, given the general obsession over breed and breed personalities, more scientific progress has not been made in this area. However, researchers are beginning to make headway.

The Smartest Breed

Even though we are reasonably sure what a Labrador retriever like Arthur will look like, we know a lot less about the way he thinks and solves problems, and how many of these cognitive skills are heavily influenced by genetics. For example, if Arthur's father and mother have excellent short-term memories, will he inherit a good memory? What about his self-control? Or his theory-of-mind abilities? And what about the other cognitive skills that predict success in service dogs?

When people list the characteristics that they think make up a great dog, being "smart" is usually one of them. No one says that they want a great dog who is cognitively unremarkable. But as we have seen at the Puppy Kindergarten, the concept of intelligence is problematic. There are many different types of intelligence, and they can operate both independently and in synchrony.

There have been claims that certain breeds are "smarter" than others. The only previous attempt to quantify intelligence in a large number of dog breeds occurred in a 1994 study by Stanley Coren.[12] Dog trainers were asked to subjectively rate which breeds of dogs were most trainable. Basically, opinions (or stereotypes) about trainability were used as a stand-in for actually measuring intelligence. Then Coren ranked each breed according to the results, with the top three breeds listed as border collies, poodles, and German shepherds. This list, despite its obvious limitations, had a lasting influence: These breeds are often considered to be the most intelligent dogs even today.

However, identifying cognitive differences in dog breeds is complicated. Behavior that seems simple can require a rich combination of cognitive skills, whereas behavior that seems complex might require no cognition at all. Cognition is always affected by genes *and* the environment, and it is no easy task to figure out the relative contribution of each. Attempts to find breed differences in cognitive performance, including one of our own studies, typically include less than a hundred dogs from a handful of breeds.[13] To determine breed differences in cognition, we would need to test thousands of dogs from a range of different breeds and ages. We would need to use experiments that we knew tested the precise cognitive skill we were looking for. Only then could we examine the heritability of certain cognitive traits—how much of the variation in cognitive skills can be explained by genetic variability. Even if we worked on this for the rest of our lives, we could never test enough dogs.

Luckily, thousands of dog owners stepped up to do it with us.

Dognition

Back at the Puppy Kindergarten, Congo loves playing games. When we turn to him and ask, "Congo, do you want to go to school?" he is always thrilled. He already seems to understand what we say, but he definitely knows the word "school," and he knows people like Margaret and Kara by name. He trots down to the kindergarten and into the classroom and looks for whomever he has heard needs him for the day.

You might think Congo's love of school is unique, but even before we started the Puppy Kindergarten, we had tested the cognition of well over a thousand dogs, and almost without exception, all of them loved playing the games. They loved coming to Duke so much that their owners would drive here from several states away. Owners told us how their dogs would start wagging their tails as soon as they realized they were coming to Duke, and once they arrived, they would drag their owners behind them to the classroom.

People begged us to teach them how to play these games at home. We realized that there were not that many games like ours for owners to play with their dogs. There were some puzzle games, and classics, like fetch, but playing our cognitive games offered a unique way to spend time with dogs and understand them better. We love playing these games with Congo because it allows us to truly engage with him. Even more than when we absent-mindedly take him for a walk or repetitively throw a Frisbee, we can really focus on the choices he makes. He gets our undivided attention. During the games, he often has us in stitches. For one thing, he is the largest dog we have had in the Puppy Kindergarten—even graduates who come back to visit are never as large as Congo. So when we pilot the puppy games, he seems to know that he is supposed to be a puppy and will try to squeeze himself into the small square where the puppies are supposed to start. Then, he tiptoes daintily across the testing mat, as though he were a tiny puppy. He reaches for the treat with his lips so he doesn't touch the equipment. And when he has had enough, he pushes the experimenter over with his nose and zooms around the room. Playing these games with Congo is just another way of getting to know who he really is.

So, in 2012, years before the Puppy Kindergarten, we launched Dognition,* an online citizen science project

---

* We do not receive any money or sponsorship from Dognition.

where dogs and dog lovers can play the same kind of games that we played with dogs who visited Duke.

Dognition is an online platform with ten cognitive games that are similar to the ones we play at the Puppy Kindergarten.[14] The games measure how well dogs understand pointing gestures, how much self-control they have, and how well they remember things and find hidden treats. Dog owners watch the instructions then play the games with their dogs, inputting the dog's choices as they go. Once a dog has finished all the different games, Dognition generates a report that compares the dog's results to all the other dogs who have played the same games.

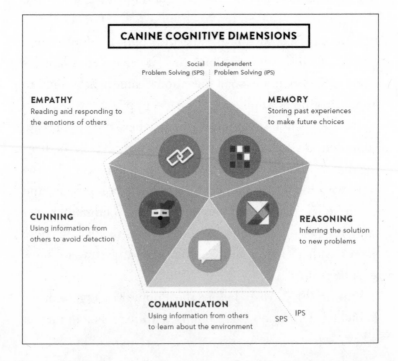

CANINE COGNITIVE DIMENSIONS

Social Problem Solving (SPS)   Independent Problem Solving (IPS)

EMPATHY
Reading and responding to the emotions of others

MEMORY
Storing past experiences to make future choices

CUNNING
Using information from others to avoid detection

REASONING
Inferring the solution to new problems

COMMUNICATION
Using information from others to learn about the environment

SPS   IPS

Around fifty thousand dog owners have signed up and contributed to Dognition. They played with their dogs in their living rooms, kitchens, and basements. People all over the world played Dognition with all types of dogs, creating the largest dataset on animal cognition in history.

These citizen scientists created a dataset that is very valuable scientifically. Because the games were based on previously published studies, we were able to compare the results from citizen scientists to conventional laboratory data. The results from dog owners mirrored the results from our research group and other research groups around the world.[15] This meant that not only did thousands of dogs and their owners have fun learning and playing new games, but the data was also usable, and we could ask questions that no single scientist could hope to answer, even in several lifetimes.*

With our citizen science data we began to examine questions about the cognitive abilities of breeds and breed groups. Before we analyzed the Dognition data, we'd thought that the cognitive performance of breeds might vary according to the function of the breed. We thought cognition was probably shaped through artificial selection to enhance a breed or breed group's ability to solve problems related to their jobs. Sporting dogs, for example, might be best at co-

* It was the Dognition data that initially falsified the use of a one-dimensional concept of intelligence, where a dog can be ranked from smart to not so smart, and showed dogs have a variety of cognitive profiles created by different combinations of skills across multiple types of intelligence.

Coren, 1998	Eye contact	Pointing	Memory
Top 3 breeds	**64 - 65 sec**	**73-76%**	**84-89%**
1. Border collie	1. Springer spaniel	1. Burmese mountain	1. Pitbull terrier
2. Poodle	2. Cocker spaniel	2. Russel terrier	2. Water spaniel
3. German shepherd	3. Water spaniel	3. Springer spaniel	3. Portuguese water dog
Bottom 3 breeds	**31 - 36 sec**	**56-57%**	**66-69%**
1. Bulldog	1. Akita	1. Shibu inu	1. King Charles spaniel
2. Basenji	2. Shibu inu	2. Greyhound	2. Havanese
3. Afghan hound	3. Greyhound	3. Staffordshire terrier	3. Western terrier

If there was one type of intelligence or a particular learning ability that explained dog cognition, we would predict the same breeds at the top and bottom of each list. The pattern we find suggests different breeds rely on different cognitive skills like eye contact, following pointing, and memory.

operative communication since they have to coordinate with humans. Working dogs might have better memories since they have to learn a variety of skills. But this is not what we found.

Evan led our team in the analysis of the performance of over seven thousand purebred dogs from seventy-four dog breeds, grouping the breeds according to the designation of the American Kennel Club. Each Dognition dog was coded based on whether they belonged to the herding, hound, toy, non-sporting, sporting, terrier, or working group. Evan and the team then compared the performances of all the groups. They found little evidence for cognitive differences between the groups of dogs. All the different breed groups performed approximately equally across nine of our ten cognitive measures.

Only one game revealed a slight but significant difference. Sporting and working breeds were more likely to follow a human pointing gesture instead of relying on their memories. In contrast, herding and terrier breeds were more likely to rely on their memories over a human pointing gesture.

Unlike Stanley Coren's study that listed border collies as the smartest dogs, our results did not point to a particular breed or breed group as somehow being the most "intelligent."[16] No one breed was consistently the top performer for each cognitive domain just as no one breed was consistently at the bottom. Our analysis also found that border collies, poodles, and German shepherds were not ranked in the top three breeds in any of the cognitive domains. Overall, the historical function of a breed, whether herding or retrieving, told us little about the cognitive skill any individual dog might have.

However, we did find some differences related to breed size. For example, Congo's brain is much larger in size than that of a dachshund. Does this mean Congo is smarter than the dachshund, or does the dachshund have a miniature brain that is just as powerful? We found that the answer depends on the cognitive skill. For the most part, that tiny dachshund brain keeps up. In six of the ten games, there were no differences between dogs with large brains and those with small brains. For example, there was no relationship between brain size and skills in the cooperative-communication games or in the physics games.[17] However, we did see differences in games that test executive function, like memory and self-control.

The Dognition memory game was similar to the game

we play at the Puppy Kindergarten, except instead of hiding the treat for only twenty seconds, owners hid the treat for up to two minutes. Larger dogs tended to remember where the treat was for longer than smaller dogs. In the Dognition version of the self-control game, owners put a treat on the ground in front of the dog and told them not to eat it. Then the owners either watched with eyes open, closed their eyes but still faced their dog, or turned their back to their dog. Again, large breeds had more self-control in resisting the temptation to take the treat. Smaller dogs were quicker to disobey. But while smaller dogs were quick to ignore their owners and take the forbidden food, they were also more strategic. Smaller dogs were more likely than larger dogs to steal food when their owner's back was turned or their owner's eyes were closed than when their owner was still watching. Smaller dogs were sensitive to these subtle cues of whether they were being watched in a way that larger, more inhibited dogs were not.[18]

How Genes Relate to Success

If the historical function of each dog breed has little connection to cognitive skills, what role do genetics play in the inheritance of the cognitive skills that are important to service dog success? In the Puppy Kindergarten, we realized that cooperative communication, self-control, memory, and understanding of physics affect service dog graduation, and our Dognition data could help us figure out the relative contribution of genes and the environment. Most animal behavior studies on heritability have focused on species like flies, fish, or mice, but these studies do not tell us much about complex cognition in dogs. Since all the dogs in Dognition played the same cognitive games and the owners specified their breed, we can see what effect genetic variation between breeds has on cognitive performance.

Evan and Gita Gnanadesikan analyzed the Dognition performance of over fifteen hundred dogs from thirty-six different breeds to test the heritability of different cognitive abilities. Gita used a massive genetic database to generate estimates for the genetic relationship among the thirty-six breeds, and then she examined how genetic similarities and differences among these breeds mapped onto measures of their self-control, communication, memory, and reasoning abilities.*

Gita found that some of the cognitive skills were highly heritable. Self-control is the highest—45 percent of the variation in performance is explained by genetic variation. In cooperative communication it is 41 percent. This is similar to the levels of heritability observed in studies of human cognition. But memory and physics were not as heritable— only 20 percent of the variance in performance in memory and physics in a breed can be attributed to genetic variation; the other 80 percent is due to other factors. Once Gita knew the relative magnitude of the genetic effect on cognitive performance of the dogs, she could then look for genes that were associated with breed differences in cognition. She found 188 different genes that were linked to performance

* She also controlled for the training experience and size of the dogs in her analysis.

on at least one of the different types of cognition measured. Each type of cognition had several dozen linked genes, and in most cases each gene probably had a small effect.[19]

However, just because you find genes associated with variation does not mean that those are the genes responsible for this variance. For instance, other researchers had found genes associated with temperament, like boldness and sociability, but previous work showed these genes were not expressed in the brain but were thought to be responsible for the size and shape of a dog's ear.[20]

But Gita found that many of the genes she identified as associated with cognitive traits were expressed in brain tissue and involved in the functioning of the nervous system.[21] There was a good chance that these genes were candidates for the differences in cognitive traits we were looking for. For example, one genetic variant, EML1, explained 29 percent of the variation in self-control across breeds. This same gene is also known to be involved in the growth and organization of neurons.[22] Gita's research tells us that there is a group of genes that plays a critical role in shaping cognition in dogs. They may also be responsible for some of the individuality we have observed across dogs. Thus, cognitive individuality in dogs is not just a result of how we raise them.

In the meantime, these discoveries might give us another way to increase the supply of service dogs. Because of high heritability, it may be possible to increase self-control in future dog populations through careful selection. In a more direct approach, we may be able to identify puppies and breeds possessing the variant of the EML1 gene, and other genes associated with strong self-control, then recruit and breed them for service dog training.

We don't focus on breed differences as part of the Puppy Kindergarten, but so many people have asked us about breed differences in cognition that we wanted to share what we have learned about breeds from Dognition—especially since we owe our discoveries to the thousands of dog owners who ran these experiments at home as citizen scientists.

All we can tell prospective puppy parents with confidence is that no one breed comes out the smartest. The appearance of a dog has relatively little to do with their cognitive profile. If any breed trait matters, it might be size. Not for all cognitive tasks, but in some cases, having a larger breed of dog might mean they have more self-control, whereas smaller dogs make up for their lack of self-control with a little more cunning.

Our research also means that the differences we observe between dogs in their ability to solve a variety of problems is not just a result of how we raise them. Skills for self-control and cooperative communication are two of the cognitive skills most important to the success of service dogs, and we have found that both are highly heritable. This same work helped us identify specific genes that are involved in building a dog's brain and shaping their cognitive performance.

These discoveries give us another potential tool in our toolbox as we continue searching for ways to improve the chances a pet or service dog will succeed with their training.

By adding these new findings to the existing process of how service dogs are carefully selected to parent the next generation, we could shape the cognition of dogs, just as generations of dog breeders have intentionally shaped their appearance. It could be a critical step toward more dogs helping more people.

But cognition is not everything and it certainly is not the only reason we fall in love with our dogs. A big part of what makes every dog special is their personality.

CHAPTER 7

A Big Personality

I f we were to list what makes Congo a great dog, the first thing that comes to mind wouldn't be his cognitive sophistication. The qualities that make Congo special are his quirky personality traits. For example, if Congo does not want to do something, he will give you a long, low bow to say, "No thank you." Or sometimes late at night, he will stand at the door and bark, as if he needs to go to the bathroom. When you come to open the door for him, he just leans against your legs and swishes his tail, which means he doesn't need to go to the bathroom at all. Instead, he just wants a hug.

Everyone who has a great dog has stories like these. But if you're a prospective puppy parent, is there a way to predict what your dog's personality will become and what factors influence it? Though the best evidence suggests breed predicts relatively little about cognition, does a dog's breed influence their personality? And as scientists, how can we even measure how personality develops in the first place?

You might have heard that Labrador retrievers are "friendly and active" with an "easygoing personality." Or that English sheepdogs are "adaptable and gentle." But the strongest selection pressure on dog breeds targets their physical appearance. Intentional and consistent selection on personality traits has been more limited. The exception may be selection on personality traits that improve a dog's success with specific jobs, like hunting or guarding. But even this selection tends to be on isolated populations within a breed. Just like with cognitive strengths and weaknesses, there is so much variation in personality within a group that it is difficult to generalize across an entire breed.

Even though Puppy Kindergarten puppies are all the same breed, their personalities couldn't be more different. If Arthur is the quintessential Lab, then his classmate Zindel is the anti-Lab. For a start, Zindel is a Labrador retriever who doesn't retrieve. Not every puppy in the kindergarten is born a retriever like Arthur, but there does seem to be a moment between ten and twelve weeks old when the ball transforms from a boring object to a *toy that can fly!*

Zindel never arrives at this revelation. There are not many puppies who get a retrieval score of zero, but Zindel is one of them. When you throw the ball for him, he gives you a look that says, "If you don't want it, why would I?"

Zindel does not like water. He views the puppy pool with displeasure and distrust. He even avoids puddles. He would never dive off a boat into the freezing North Atlantic water to retrieve fish. Not in a million years.

His performance in the cognitive games is erratic. One week he scores 100 percent, correctly answering every trial. The next week on the same game he gets every answer wrong. The next week he is at 50 percent, as if he is closing his eyes and choosing at random. He does not improve or get worse over time. It is impossible to fathom what kind of cognitive development is occurring—something *is* happening, but it is very confusing to watch.

Arthur, true to a Labrador's reputation, is friendly and outgoing. Arthur loves people, is good with children, and is always up for a game of tug or chase or tag. Arthur doesn't care who wins, he just loves to play.

Zindel is moody. One minute he bursts into the puppy park, ready to be the life of the party. The next minute he is sulking under a bench with a look on his face that says, "Whatever you want, the answer is no."

Zindel is also a sore loser. He quits any game he is not winning. He will often hoard toys, dragging them under his bench and then lying on them like a dragon on a pile of jewels. If he gets in a mood, he picks on the smaller puppies. Then, just when you've made up your mind that Zindel is the least likable puppy you've ever met, he will lie on his back and let the smallest puppies jump on his chest. Or he will logroll across the entire length of the puppy park, right into your lap. The differences between Arthur and Zindel are fascinating. But how would you go about measuring them scientifically? Will these differences stay consistent as they grow up? What kind of personality makes the best service dog? And how has our research changed the way we recommend people choose a great dog?

The Long Shadow of Temperament

At the Puppy Kindergarten, we do not study personality. Instead, we study temperament. Temperament is an important component of personality. Although your personality can change and grow, your temperament is with you from the start. Temperament, like all behavioral traits, has a genetic component, meaning that your genes influence your temperament, but do not control it. Temperament shows consistency throughout time. You might learn to control your temper, but you will always have one. As one researcher described it, your personality is a symphony, and your temperament is the key in which the symphony is played.[1]

Jerome Kagan, a pioneering temperament researcher, tested the temperament of human babies using a single criteria—how babies respond to something new.[2] Kagan showed four-month-old babies a range of new objects and toys. Around 40 percent of the babies were calm and curious. These babies tended to grow up to be emotionally spontaneous and quick to make friends. Around 20 percent arched their backs and cried. These babies tended to grow into eleven-year-olds who were shy and hesitant in new places. As teenagers, these same children were more likely to be anxious and subdued.

Temperament and cognition are usually treated as two separate processes. Temperament is the probability that you will have a specific emotional reaction to certain situations.[3] Cognition is the way your brain receives, processes, and uses information to solve problems. Temperament includes how you behave in situations that are discrepant, unusual, or unfamiliar, for instance when you encounter something strange and new. Cognition is how you use insight or inferences to make decisions and react flexibly to solve problems. But in reality, temperament and cognition do not exist as two distinct processes in the brain. Self-control can be classified as both a temperament trait and a cognitive trait. Temperament researchers call it "self-regulation," while cognitive researchers call it "executive function."[4] Self-control is instrumental in problem-solving, but it is also crucial in regulating behavior in emotionally charged situations.

Temperament and cognition interact in complex ways, and there are traits like self-control that straddle the temperamental and cognitive domains. Your personality is a combination of your temperament and cognition, and how they interact with the environment as you grow.

This is why, at the Puppy Kindergarten, the games we play are not all about problem-solving. Temperament is also an important part of what makes a great dog. And because temperament appears at a very young age, puppies are the perfect place to start.

Something New

A classic temperament test is to see how someone young reacts to something new and unexpected. To an eight-week-old puppy, there is nothing newer or more unexpected than Kitty.

Kitty is a life-sized robot cat. She is a tabby ginger with a pink nose and giant eyeballs. Kitty is programmed to follow a certain sequence of actions. First, Kitty sings an introductory song and waves her paws in the air. From then on, Kitty responds to movement. She can stand up and do a commando crawl. She purrs a lot.

Arthur is the first of the class of Spring '20 to meet Kitty. At eight weeks old, he already has impeccable manners. He approaches Kitty to introduce himself and politely sniffs her bottom. She purrs. It is clear that Kitty does not smell like anything Arthur has encountered in his short life.

Kitty is the puppy version of the novel object test that child psychologists use with babies. When an animal encounters something strange and unusual like Kitty, you can note how fast and how close they approach, whether their behavior is friendly or antagonistic, and how long they stay in proximity.

Arthur makes a wide circle around Kitty while she meows and waves her paws around. Arthur sits down and observes her. He waits for a respectable period—almost to the end of the ninety-second trial. Then he puts his paws on the pen and asks Kara if he can please go back to playing with his friends in the puppy park.

Two weeks later, Arthur meets Blue, a turquoise robot triceratops with a magnificent crest. Although Blue is a dinosaur, she has body language that Arthur understands. She play bows and wags her tail. She cocks her head to one side. Arthur, always the gentleman, walks over to introduce himself and sniffs her bottom. Blue stands on her hind legs and says, "Wow!"

Arthur is still unsure about robots, but he is two weeks older than last time and has gained some confidence. He does not ask to leave. He retreats, but stays closer than last time, while Blue gurgles and talks to herself. Arthur sits quietly and watches Blue for the rest of the trial, as though he

is the one conducting the experiment and Blue is the one under observation.

Two weeks later Arthur meets Lion. It is unclear what the toymakers had in mind when they programmed Lion. When Lion "sees" Arthur, he gives a stuttering roar.

Arthur walks over to say hello, and Lion spins around in a frantic, jerky circle, as though he is feeling ill. Arthur does not judge him. He treats Lion as respectfully as he has the other robots and tries to gently sniff Lion's bottom, but Lion is spinning so fast, Arthur can't put his nose in the right place. Arthur does the best he can, but then resumes his role as a researcher, studying robots in their natural habitat.

Arthur has what child psychologists would call an easy or nonreactive temperament. He is friendly and curious, a little cautious but unafraid, and gains confidence as he matures. Zindel, on the other hand, reacts differently every time we introduce a new robot. When he meets Kitty, he runs away crying.

Two weeks later, he is friendly and charming to Blue the

triceratops. Another two weeks and he is sulky and bored, hardly acknowledging Lion's existence.

In Kagan's longitudinal study, 40 percent of the babies did not neatly fall into any category. They might be anxious in some situations and not others. Or they might be difficult with some people but easy with others.

Babies on either end of the spectrum—difficult, reactive introverts and easy, nonreactive extroverts—were closely followed into adulthood. Those who were not easily categorized tended to fade out of the analysis. After all, if you can't fit someone into a category, it is difficult to compare them to others and see how they change over time.

Zindel is one of these nonconformists. And while everyone falls in love with Arthur's easy manner and predictable adorability, there is something exciting about Zindel. He is like six puppies for the price of one. Zindel's contrasting behavior is yet another example of why attributing temperamental traits to different breeds is problematic. While there has likely been more intentional selection on dog temperament than cognition, there is still significant temperamental variation within any breed. Arthur might have a classic Labrador temperament, but what do we make of Zindel, whose temperament we can't categorize even in the broadest sense? This does not mean that there are no similarities in temperament within a breed, or that you could not create them. In fact, Canine Companions, through careful selection for friendliness, have created a common temperament response to strangers in their population of dogs.[5]

Because while Arthur and Zindel might differ in the way they respond to meeting new robots, they respond the same positive way when meeting new people.

Stranger Danger

After a satisfying morning of games and treats, Kara puts Arthur on his leash and together they walk into the waiting room. There, a stranger in a yellow rain jacket and ski mask

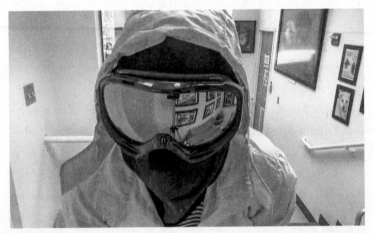

The stranger in the waiting room . . .

sits on the stairs. They are talking on the phone and they sound frustrated.

"Come on, ref—that was clearly a charge, not a blocking foul!"

The stranger suddenly stands up. They lean on their cane to help them walk and head straight for Arthur.

Arthur watches the stranger and wags his tail with uncertainty. The stranger gets louder.

"Aww, put on your glasses!"

The stranger waves their cane right at Arthur. He backs up toward Kara and yawns at her feet. He presses against Kara's legs, a puppy version of a hug.

The stranger sits down, removes their hood to reveal their face, and says, "Hi, puppy!"

This is the key moment. At just eight weeks old, before he has seen much of the world, or experienced anything as nerve-racking as a tall stranger swinging a large stick, Arthur walks right over to the stranger, tail wagging. He sniffs, sits down, and does a little hop on his bottom. Then he jumps up, puts his two front paws on the stranger's lap, and tries to lick their face.

Zindel's turn. At eight weeks old, he struts out of the classroom, having just aced the odor test. Full of confidence, Zindel sees the stranger, and he does not hug Kara for comfort. Instead, he turns and ignores the stranger, who keeps approaching. When the stranger starts waving their big, heavy stick around, Zindel tries to slip past them.

But then the stranger pushes back their hood and says, "Hi, puppy!"

Given Zindel's response to Kitty, you might expect him

to run for his life. Zindel does the opposite. All fears are forgotten, he just about wags the tail off his bottom and tries to leap into the stranger's lap.

Although Arthur and Zindel vary in how they feel about animate robots, they both love people. All Canine Companions dogs do. They are not only people lovers who are unafraid of strangers, they are actually *attracted* to strangers.[6]

Our dogs' ability to act as alarm systems has been invaluable over the years. In the Peloponnesian War, guard dogs warned the Corinthians of an attack from the Greeks. In 55 B.C., giant dogs known as mastiffs defended the island of Britain against Julius Caesar. During the Napoleonic wars, a black poodle called Moustache alerted French troops to an Austrian spy in their midst. However, as we move into more densely populated urban spaces, and bring our dogs on vacations and out into public spaces, aggression or even wariness toward strangers is less desirable. For decades, service dog organizations have been heavily selecting dogs who are attracted to strangers, and the result is a dog who is happy anywhere, in a playground full of children, at a busy café, or under a desk at work.[7] For people who want their dog to share more of their lives, it is essential they choose a dog who is open to meeting strangers.

Impossible Tasks

Another temperament trait that Arthur, Zindel, and other service dogs have in common is that they tend to make more eye contact than other dogs. Eye contact is important because it stimulates oxytocin.* Oxytocin is a tiny neurohormone that is involved in almost every aspect of our social lives, particularly when it comes to bonding between parents and their babies. When parents stare at their babies, it triggers oxytocin release in the babies, which causes the babies to stare back at their parents, which makes parents want to stare more at their babies, and so on.[8]

We know oxytocin is important to our relationship with dogs because dogs have managed to hijack the oxytocin loop.[9] Have you ever noticed your puppy staring at you for no reason? When your puppy stares at you, you experience an increase in oxytocin, which makes you want to stare more at your puppy. Your puppy also experiences an increase in their oxytocin, which makes them want to continue staring at you. It's like they are hugging you with their eyes.†

Eye contact can be extremely important for puppies, and for dogs in general. People with dogs who make longer eye contact show a higher increase in oxytocin than people whose dogs make eye contact for only a short amount of time.[10] In one study, dogs who made more eye contact (with the whites of their eyes visible) at shelters got adopted

---

* Oxytocin is transmitted from the brain directly into the bloodstream and along nerve fibers to the nervous system. It is what makes you feel good when you are touched by a loved one, get a massage, or enjoy a good meal. Oxytocin has pain relieving properties and can also decrease stress and blood pressure.[11]

† Social interactions with people also are thought to affect the oxytocin levels of dogs. Experiments suggest dogs have an increase in oxytocin after they have a friendly interaction with a person.[12] People also have decreased measures of stress and increases in oxytocin after hugging a dog.[13]

faster.[14] People with dogs who make more eye contact report being happier with their dogs.[15]

So how do we measure eye contact in puppies? Allow Westley to demonstrate.

Westley has more face wrinkles than seem possible. They make him look worried, as though he wakes up in the morning and reads every catastrophe in the news. He also has a fairly slow way about him, which means every time he approaches it's like he has something awful to tell you.

To see how long Westley makes eye contact, Kara gives him an "impossible task." Kara drops a treat in a plastic container. Westley is very food motivated, so as soon as Kara gives him the word, he waddles over and eats it. Then Kara makes the test harder—she puts the treat in the container but puts a lid over it. Westley still has no problem. But then, on the next try, Kara does something really frustrating. She snaps the container shut. It is one of those snap-lock containers that no one without opposable thumbs can get into.

Westley and the impossible task

When you are faced with an impossible task, you have three choices. You can give up, you can keep trying on your own, or you can ask for help. We have found that military dogs are relentless; they *know* they could get the treat a minute ago. They paw the container and bite on it. They never give up. But they rarely ask for help.

Westley is different. When the container doesn't open, he stares at it for a few seconds, then almost immediately looks at Kara. He stares deeply into her eyes. After sixty seconds, Kara opens the container. Westley eats the treat and then gives Kara a kiss.

Westley has the distinction of making the most eye contact of everyone in the class of Spring '20. But even Arthur and Zindel make more eye contact than the average dog, and they make way more eye contact than other Labrador retrievers, like military dogs.

Because eye contact can be so important to the relationship between people and their dogs, what can you do with your puppy at home to help them make more eye contact with you? We have found that a tiny bit of early experience can lead to an increase in the amount of eye contact a puppy makes with you.

Puppies who play the impossible task every two weeks, between eight and twenty weeks of age, make about twice as much eye contact as puppies who play the game for their first time at twenty weeks. Just playing this little game, which takes about five minutes once every two weeks, leads to a big increase in eye contact.

If shelters could play this eye contact game with their dogs, it might improve the rate of adoptions.[16] Dog owners playing this game at home with their dogs could make their relationships even better. If you were tempted to try any of our games with your puppy, the impossible task should probably be it.

The Future of Breeds

As our lifestyles have evolved, we've begun to expect different behaviors from our dogs than we used to. As discussed in the last two chapters, a puppy's breed does not tell you as much as you might expect, except what they will look like when they grow older. But what if we started to consistently select for temperament rather than appearance?

Today, dogs rarely fulfill traditional roles as guards, hunters, and herders. The very traits selected for in the past are unsuitable for our increasingly urban lifestyles. For example, barking is now considered a nuisance behavior, and stranger-directed aggression is one of the leading causes of serious injuries caused by dogs. A dog who guards the house might be convenient, but not at the cost of attacking everyone who comes to the front door.

Our research with service dogs and military dogs has shown us that directly selecting for temperament can have drastic effects. Canine Companions has consistently selected for dogs with calm, easy temperaments who are attracted to strangers and make a lot of eye contact. The selection for a friendly temperament in Canine Companions dogs has been so strong, it has even changed their physiology.

When Evan measured the oxytocin in pet dogs and Canine Companions dogs, he found that Canine Companions dogs have a higher circulating level of oxytocin and a higher amount of total oxytocin compared to pet dogs.[17] This shows that it is possible to create a breed of dog that we can claim with reasonable certainty "is friendly, easygoing, and good with children." The key to doing this is to shift the focus from what a dog looks like to the kind of dog you want to live with.

Most likely, this is a dog who can be part of your family. Who welcomes guests into your home and is at ease with your friends. Who can accompany you on hikes and neighborhood strolls, and the occasional well-deserved vacation.

The exceptionally calm and friendly temperaments of Canine Companions dogs like Westley, Arthur, and Zindel demonstrate the effect a careful breeding program can have on a population of dogs. This could be the future of breeds. Who a dog is at heart is at least as important as what they look like.

What Kind of Puppy to Get

Even though a dog is much more than its breed identity, many people will still obsess over breeds. It will take time to convince enough people to emphasize selection for temperament—rather than short legs or flat faces—and even more time for these changes to appear. In the meantime, we get many requests for advice on what breed of puppy to choose. Our first recommendation for anyone who wants a dog is to foster one. Many shelters and nonprofits have foster programs, where you take a dog home for either a set time or until adoption.

Foster dogs are usually older than puppies (although puppies can be fostered too), so their personalities are more settled, and you can see if they are a good fit for your family. Having a dog stay with you gives you the chance to get to know each other, test your chemistry, and see what they will be like in a home environment.

At worst, you have given a dog a change of scenery from a shelter and gotten some experience caring for a dog. At best, you will become the new forever home for a dog that would otherwise be homeless.

If you can't foster, our second recommendation is to go to a rescue organization, like a shelter. Talk to the staff about the kind of dog you are looking for and if they have a dog in mind, ask them what they know about the dog's history and any behavioral or health records. Rescue organizations usually have a dog who has been there for a while, who they know well, and who would be perfect for the right home, but for some reason, has not been adopted yet. Rescue staff have an interest in being as honest as possible—they don't want a dog to be returned the next day—and they are usually more than willing to think with you about who might be a good fit.

If you really, really want a puppy, rather than an adult dog, our final recommendation is to consider a rescue from a shelter or foster home first. If you are set on a purebred puppy, you can also check breed rescue organizations.*

You and your new dog will probably spend the next decade together. It is worth being careful about who you choose and using your choice to contribute to the health and welfare of the next generation.

But in the end, if your puppy is a mystery puppy, and you have no idea what breed they are or where they came from, don't worry—genes aren't everything.

How you raise a puppy is just as important as where they come from.

---

* You might have heard that mixed breed dogs are less likely to have genetic health problems due to a phenomenon called "hybrid vigor"— where the offspring of two purebred strains are healthier than the parents. However, there is a lot of variation in mixed breed dogs, depending on what mix they are and whether there is any genetic disease in their heritage. There is some evidence that mixed breed dogs are healthier than purebreds. This includes a lower rate of inherited diseases like skin cancer, hip dysplasia, diabetes, eye disorders, and epilepsy.[18] Several studies found that mixed breed dogs live on average one to two years longer than purebred dogs and experience a slower aging process.[19]

Class of Spring '20, left to right: Zindel, Arthur, Wisdom, Aurora, Westley, Zola, and Yonder

CHAPTER 8

Critical Experience

Given that all puppies are individuals and that genes play a role in their cognition and temperament, is there a certain environment that gives a puppy the best chance of becoming a great dog?

Should puppies meet as many different people as possible, or will a crowd overwhelm them? Should puppies meet other dogs as soon as they can, or wait until after they are vaccinated? How much time should puppies spend with other puppies of the same age? What types of places should a puppy go and where should they avoid? To what degree can experiences shape a puppy's intelligence while their brain is growing so rapidly? These are just a few of the many questions people have when raising their puppies, but clear answers can be hard to find since there has been limited scientific research. Drawing on individual experiences and observations can lead veterinarians, breeders, and trainers to have different or even conflicting opinions. In this chapter, we share why the critical experiences your puppy needs most are provided by some very conventional rearing methods. We arrive at this conclusion using our experimental design that allows us to carefully compare our kindergarten puppies to those raised in a family home.

We know the social world is crucial to cognition, temperament, and behavior because of what happens to animals who grow up without it. Baby animals need care from their mothers, interactions with others of their kind, and rich experiences with the environment they must learn to negotiate.* In many species, ranging from monkeys to mice, early deprivation and isolation have been associated with a range of neurophysiological disorders and behavioral abnormalities later in life. Puppies are unique in that they not only have to learn how to interact with their own species, they need to bond with an entirely different one. One consequence of domestication is the evolution of an expanded socialization window that gives puppies a longer time to form bonds.

In wolf puppies, the socialization window is relatively short—from two to six weeks old.[1] This is the period where wolf puppies must be intensively socialized with humans if you want them to tolerate humans as adult wolves. A human must bottle-feed them, sleep with them, and be their pri-

---

* In mammals, the socialization window is a period of reduced fear and reactivity. It is the time when young mammals are developing their senses and becoming familiar with others of their species and situations they might encounter later in life.

mary source of contact. Even with this intense socialization, wolf puppies never quite bond with their person like dogs do, and they are still wary and afraid of other people.

Dog puppies are characterized as having a longer socialization window,[2] and they quickly bond with humans during this period. Because the socialization window in puppies is longer, it offers us an opportunity to look at what environmental factors most influence cognitive development. The best way to scientifically identify these environmental factors is to experimentally compare puppies who have been raised in two different ways.

Our first puppy group is called the "home puppies" because they are raised in the homes of puppy raisers, special people who have volunteered to raise a service dog puppy—not for themselves but for others. Puppy raisers take their puppy to classes, provide progress reports, and consult with experts as they complete their puppy's earliest stages of training. Puppy raisers introduce their puppy to other dogs and dozens of people, travel to different places like shops and other homes, even airports or hospitals. They may have another pet dog at home but no other puppies.

Our second group of puppies, called the "kindergarten puppies," is raised at our kindergarten. It took some serious thought to figure out how to give them an even richer environment than puppy raisers provide. Instead of being raised by just one family, the puppies are raised by dozens of students. The puppies travel all over Duke's campus, to the library, the hospital, and the dining hall. And instead of growing up with just one other dog, or without any dogs at all, the puppies grow up with other puppies. They play, wrestle, romp, and sometimes fight puppies of their own age.

To test the impact of these different approaches to rearing, we compare cognition of the home puppies to the kindergarten puppies. The main difference is the home puppies come in every two weeks and play the same cognitive games as the kindergarten puppies. In the end, we will see which approach to rearing increases the chances of success for our puppies. But it will also reveal which type of upbringing is best for your puppy, since home puppies are raised similarly to the way you raise your puppy.

To get a sense for how different the early experiences of our kindergarten puppies are from the home-reared pups, you really have to live it for a day.

A Day in the Life of Wisdom

Wisdom has three dorm parents, who each take care of her for a few nights a week. In the morning, Wisdom wakes up and stretches, and immediately her dorm parent takes her out to the bathroom. She comes back inside, and if there is time, she might snuggle down and snooze in her dorm parent's lap. She has a relaxing breakfast, gets dressed in her cape and doggy leader, and walks with her dorm parent to school.

The Duke students are not usually up as early as Wisdom, but if they are, they will stop and hug her and wish her a good day at school. At school, Wisdom rushes in to see her classmates. The dorm parent hands the leash to one of the morning volunteers, Wisdom joins her classmates, and everyone goes for a walk around Duke Pond.

Duke Pond is a full sensory experience. Wisdom sees ducks paddling in the water, bees buzzing through the wildflowers; she feels the dew on her whiskers, the dirt under her paws; she smells the rabbits and raccoons who are hiding in the woods. After the walk, her morning volunteers take her and the other puppies to the bedroom. They take off her collar and give her a kiss as she settles down for her nap. When she wakes up a few hours later, new volunteers are there. They hug her hello, take her outside to the bathroom, and then comes one of her favorite times of the day—lunch.

After lunch there is training, which Wisdom adores. She has done her training homework and is set to crush her midterms. New volunteers come and it is time for Wisdom's other favorite time of the day: recess. Recess is in the puppy park with her brother Westley and all her friends. They play in the puppy pool and on the puppy slide, chase one another, and chew on an endless rotation of toys that are switched out each day.

In the afternoon, Wisdom goes to class. The valedictorian of the class of Spring '20, she excels in almost every cognitive game out of the gate. At eight weeks old, she scores 100 percent on the marker test, and unlike some puppies who start high then get worse, Wisdom hits her ceiling and stays there. At ten weeks old, she has perfect recall in the memory games. By twelve weeks, she masters the self-control test that Ying did not start to understand until he had almost graduated.

Wisdom consistently stays top of the class in physics. Her classmate Arthur is the most enthusiastic and dedicated

player in the retrieval games, but at ten weeks old, Wisdom gets the highest retrieval score possible (which Arthur can't quite manage). Then, perhaps because she is a good friend, she lets Arthur overtake her. In the temperament tests, Wisdom is playful with Kitty, Blue, and Lion. Wisdom is nonplussed at the stranger in the hoodie who yells at the referee and waves a big stick in her direction and happily goes over to be petted when the tantrum is over. She plays well with others but does not mind being alone.

Sometimes, Wisdom puts on her cape, jumps into the golf cart, and rides to the hospital. Wisdom walks across the shiny floors of the hospital lobby, among people who are sick, afraid, or relieved. She is accompanied by Congo, who walks calmly beside her wearing his hospital badge, so everyone knows that they are allowed. They take the elevator to the second floor that smells of disinfectant and together walk down the hall to a small room where the nurses of pediatric cardiology are waiting. Sometimes they are upbeat and happy to see her. Other times they are so sad, they cry when they hold her.

On other days, Wisdom visits the wellness center with its water feature and singing bowl. She meets dozens of students in a room that smells like aromatherapy oils. Wisdom navigates a crush of students who barely notice her in the cafeteria at lunch. She stands in front of an audience of five hundred who are riveted by her during orientation. But the key component is that even while meeting thousands of people in the space of a few months, all the experiences are positive.

Wisdom is never shouted at. She is never punished in any way. If she wants attention, there is someone to play with her or cuddle her. If she wants to be left alone, her space is respected. And at the end of the day, her dorm parent picks her up from kindergarten and they walk home together. After her dinner, if she is still feeling social, she will visit the dorm common room, where there is always someone who is glad to see her. When the day is done, she curls up in her crate under the desk of her dorm volunteer and falls asleep.

When people visit the kindergarten and see the puppies lying on their backs and having their paws massaged (this makes it easier to clip their nails), their faces wiped, and their fur brushed in long, loving strokes, they shake their heads in wonder. It is amazing how much love and attention these puppies get in the ten weeks they are in the kindergarten. The question is: Does it make a difference?

By now, you might be starting to wonder if you need to quit your job, recruit a hundred undergraduates, and bring home another six puppies to re-create Wisdom's experiences.

The good news is, according to our current results, home puppies did not have significantly different outcomes than dogs raised in the kindergarten. To compare the two groups, we played the same games with both groups of dogs—ones that assessed their cognitive and temperamental abilities— every two weeks while the puppies were between eight and eighteen weeks of age. Most of them were tested in the same room at Duke and all of them were from the same population of Canine Companions service dogs. Despite the different ways we raised the two groups our comparison revealed that both groups performed similarly on all the cognitive tests. Neither have we seen evidence of major differences in their temperament tests. So far, puppies who were primarily raised in the Puppy Kindergarten do not have a higher rate of graduation as service dogs than dogs who went home to a family every day. One group was not calmer, more social, or more cognitively sophisticated than the other. It seems that the rapid brain growth that occurs during this period follows a relatively stable maturational trajectory and is not influenced by the different rearing approaches we compared.

Our findings do not imply that rearing does not matter. It absolutely does. The impacts of *not* socializing a puppy, or mistreating or neglecting a puppy, are swift and severe. There seems to be a socialization threshold—a certain amount of exposure to new dogs, people, and places that a puppy needs to thrive. Anything above that threshold may not make a noticeable difference for most dogs. And for the most part, being part of a kind, loving family is enough for a dog to turn out great.

Different Mothering Styles

What about a puppy's early life before they are weaned, when they are still with their mom and littermates? Is that when different experiences might matter more? Emily conducted one of the only studies to investigate the impact of early rearing experiences.[3, 4] She found that differences between early maternal experiences do matter. Emily studied several groups of puppies who were set to go either into service dog training or guide dog training.* She studied the puppies while they were still with their mothers, before they were weaned at six to eight weeks old. The puppies and their moms were settled in a room with a kiddie pool lined

* Canine Companions trains many types of service dogs but does not train guide dogs for the blind.

with towels. Kiddie pools are perfect for a nursing mother and her puppies. Mom can recline against the edges when she is tired, and the puppies feel safe and enclosed. Most important, the sides of the pool are just high enough so that the puppies cannot climb out, but their mom can come and go as she pleases.

Emily measured the amount of time the moms were in the pool with their puppies and observed how each mom behaved when she was with her puppies. She noticed two different mothering styles: the helicopter mom and the free-range mom. At one extreme, the helicopter mom preferred to always have her puppies in her sight and, even better, in constant contact. The helicopter mom was in the pool almost all the time. She would constantly lick and groom her puppies, nuzzling them and warming their bodies as they slept. And when her puppies were hungry, she obligingly lay down on her side, so her puppies could lie down and nurse.

A helicopter mom, right inside the pool, lying down so her puppies can nurse

The free-range mom was the other extreme—she was much more likely to leave her puppies in the pool by themselves. The free-range mom was slow to respond when her puppies whined and spent much less time in contact with them. And when her puppies wanted to nurse, she would make them work for it by standing up, forcing them to climb over one another and jump for her nipple!

A free-range mom, not only leaving her puppies alone, but taking a nap outside the pool

Emily followed these puppies for two years—long enough to see if they graduated as service dogs. Previous research predicted that the puppies of the helicopter moms should do better. In several studies, puppies with attentive mothers were less fearful and more likely to engage with their social and physical world; these puppies also had lower levels of aggression, anxiety, and fear.[5, 6] Emily's results were in accordance with these previous studies, finding that puppies of helicopter moms were more likely to graduate than puppies of free-range mothers. Puppies of helicopter moms were also less fearful of strangers—although they were more prone to separation-related behavior problems in adolescence.

But all these puppies were Canine Companions service dogs. Guide dogs are a different kind of service dog and they require specialized training (for instance, if their owner is about to do something dangerous, like walk into traffic, a guide dog has to be selectively disobedient). When Emily tested the guide dog puppies, she found that puppies of the free-range moms had the greatest chance of graduating. These puppies grew into calm, even-tempered adults, who performed better at the cognitive games. Puppies of the helicopter moms were more agitated when they were alone, took longer to solve problems, and were more reactive when they encountered the novel object task. These puppies were more likely to be released from the training program.

Emily concluded that different mothering styles are more suitable to different careers. Helicopter moms help their puppies become more secure and bonded, as well as more comfortable around people. Free-range mothering probably encourages traits like independence that are important to guide dog work.

A different study focused on other kinds of working dogs, including guide dogs, and looked at intensive socialization with people from a very early age. This intensive contact consisted of daily handling exercises, or "gentling." When the puppies were three days old, a person picked up the puppy and tickled between their toes. They held the puppy in odd positions—very carefully and only for a few seconds—for example, upside down by the paws. Or they put the puppy on a cold, damp cloth. These and similar exercises were briefly repeated every day until the puppy was around three weeks old. Handling puppies at this age is somewhat risky because they are so fragile and helpless. But researchers found that puppies exposed to gentling techniques had stronger heartbeats, improved heart rates, and a higher resistance to stress and disease.[7]

All this work suggests that the same maternal style can have different long-term impacts on each dog in relation to their future working-dog career. We hope our team's research highlights the kinds of critical experiences puppies require after they leave their moms. Puppies need to meet and positively interact with different dogs and people. Puppies who are socialized in this way tend to be more socially tolerant and confident when they are adults. But there is no real advantage to introducing your puppy to thousands of people, raising them with same-aged puppies, or scheduling intensive extracurricular activities for them like we do at Duke. Instead, you should definitely send your puppy to school—puppy school. Not only do puppy classes teach your puppy basic skills, they also teach your puppy how to have positive interactions with new people and new dogs. Introduce your eight- to eighteen-week-old puppy to your family and friends, including friendly dogs, take them out on excursions, and make sure they get plenty of exercise and lots of love.

And that will be enough.

CHAPTER 9

_____

Bringing the Puppies Home

I n the spring of 2020, when the pandemic first hit home, we, like everybody else, hoped everything might get back to normal in a few weeks. Then, as the months went by, we learned that normal might not arrive for a long time.

As summer arrives, it is clear we are in trouble. We are behind on our research. Covid infections are not subsiding and there can be no puppy class of Fall '20. Airlines are not flying dogs, and Canine Companions has paused their breeding program. Gone are the one hundred volunteers who monitor every movement of the puppies and cater to their every whim. Margaret, our vet, is no longer on campus to examine every scratch and scrape. Expert trainers from Canine Companions can't fly to Duke to help us keep the puppy training on track. We are stuck. How can we have a Puppy Kindergarten without puppies?

But Canine Companions calls, saying they have four

puppies for us to raise. So we move the Puppy Kindergarten to our house and take two puppies; our graduate student takes one puppy; and the lab coordinator takes the fourth. Fall 2020 becomes a puppy class of four, which is less than our goal but better than nothing.

As soon as we decide we should turn our house into a Puppy Kindergarten, we realize how unsuitable our house is to be a Puppy Kindergarten. At Duke, the puppies have 8,600 acres of manicured grounds, with wide sidewalks and few cars. The puppies have Duke Pond to stroll around when the weather is good and miles of indoor corridors to walk when the weather is bad.

Our house is on a busy road that is perilous for animals and children.

Our yard is a muddy mess of plants that are poisonous to dogs. The house was built on a slope so all the water drains into our backyard, creating a breeding ground for mosquitoes the size of small birds. The inside of our house is even worse. The Puppy Kindergarten was built specifically with puppies in mind—the floors, walls, and furniture can be wiped down and cleaned. There is no carpet. The puppies are never left unsupervised, but in the unusual case they are left alone for a few minutes, we can be reasonably sure they will not hurt themselves. Our house, on the other hand, is a disaster. It is insane how many objects of danger are at puppy height in a normal house—electric outlets, stairs, household chemicals.

But the real problem is online school. At the same time we are re-creating the Puppy Kindergarten in our home, we have a human kindergarten unfolding in the middle of the living room. Our pod family has two children, which means we often have four children remote learning in four different grades with all the associated chaos—iPads, laptops, and the rat's nest of cables that must be untangled daily. Pencil sharpeners under the dining table. Erasers in the air vents. We can't cross the length of a room without stepping on a Lego. Only Congo, with his cognitive sophistication and advanced level of training, is able to negotiate this disaster zone.

After spending a few days immobilized by panic, we get to work. Although a highly specific situation, our preparations can be used for any procrastinating puppy parents who find themselves unprepared for their puppy's arrival. We separate the downstairs from the upstairs with a baby gate. Every item below waist level that is not directly relevant to survival is moved upstairs. Since we have found that puppies love to go to the bathroom on soft, rectangular objects, we roll up the doormats and the living room rugs. We remove the end tables. We put a basket of old towels and pee cleaner by the door. We unplug the floor lamps and seal the electrical outlets. We move all the school supplies to the narrow safe zone of a bookshelf: short enough for a child to reach while standing on tiptoes but too tall for a puppy standing on their back legs to grab with their mouths. We sit the children down and tell them horrid stories of stuffies who were dismembered and Lego masterpieces that were mauled beyond recognition because neglectful children *brought them downstairs into the jaws of death.*

Once the living room looks like no one lives in it, we organize the containment areas. First we bring two sleeping crates with sliding trays from the kindergarten. The next piece of essential equipment is the x-pen, a foldable plastic fence that creates an instant playpen. X-pens come with connectors so you can make them into a small square or a room within a room. You can put up an x-pen anywhere in minutes. We use them for testing in the classroom, for mealtimes, and for any situation where we need the puppies to be contained for a few moments. Once the containment areas are set up, we gather the supplies. It is a fairly long list, and with every class of puppies, items are added and removed. However, the most up-to-date version at the time of writing is in Appendix I.

Outside, we enclose the flattest part of the yard, an area twenty feet wide by twenty feet long, with four-foot-high poultry fencing. The mulch is in an advanced state of decomposition, which means it is relatively soft, without any splinters. Puppies like to relax off the ground, especially when it is damp, so we put in a large dog bed that is woven fabric on stilts, and a wooden platform that used to be a wooden pallet.

Nearby, we have a metal, waterproof bin with their dog toys. We order about a thousand rolls of poop bags made from recycled, compostable material. We are as ready as we can be. Finally, on October 7, 2020, Rainbow and Sassy arrive.

Sleep and Its Cycles

Of all the milestones the puppies must pass, the one we watch most closely is sleeping through the night. A puppy who sleeps through the night is a puppy with regular bowel movements and solid stool. A puppy who sleeps through the night is a well-adjusted puppy, who feels calm and comfortable in their environment. Most important, a puppy who sleeps through the night does not wake up to cranky, exhausted parents. Our bedtime rituals at the Puppy Kindergarten are the same rituals recommended for children.

First rule, start to settle down early. At Duke, the dorm parents start the routine four hours before bedtime, at seven P.M., when they feed the puppies dinner. Around half an hour later, the volunteers take the puppies outside to poop, and the goal is that all puppies should poop sometime in the hour after dinner. Three hours before bedtime, at eight P.M., the dorm parents take up the puppies' water. The dorm parents brush their puppies' teeth, groom them, and cradle them, and by this time, most of the puppies start to doze off. After a few more toilet breaks, there is a long toilet break at ten forty-five P.M. This is a full fifteen-minute break that is the last chance for puppies to go to the bathroom. Some of the puppies are so sleepy that they literally fall asleep standing up. If this happens, the volunteers carry those puppies farther away, and the puppies must walk across the dewy grass to get back to their warm beds, which is usually enough to entice them to pee one last time.

Second rule, the bedroom must be a comfortable place that is only for sleeping, not any other activities. All Canine Companions puppies sleep and nap in crates. For several weeks, we supervise their crate time, sometimes with a baby monitor, to make sure the puppies are calm and happy in their crates, without any signs of distress. After the first few

Sassy and our daughter, settling down early

weeks, we encourage puppies to sleep in crates for the simple reason that we can leave them unattended and know they will be there when we get back. A puppy who sleeps in a crate cannot wander around at night and poop under the sofa. A puppy who naps in a crate cannot get up while we aren't paying attention and chew through electric cables. A crate is peace of mind, for us and the puppy. It is where they learn to be alone, a private space they can retreat to that is warm and safe, like a den. Their time in the crate is limited—a two-hour nap in the morning and another in the afternoon, followed by eight hours overnight (unless the puppies choose to go into their crates, which sometimes they do). The puppies sleep with no distractions—that means no toys and no blankets.

Third rule, keep bedtime consistent. Lights are out at exactly eleven P.M. Puppies have a circadian rhythm similar to ours—they are active during the day and sleep at night. But research suggests their sleep architecture is different.

We have sleep cycles of ninety minutes, and the sleep cycles of dogs are around thirty minutes. While we drift from one sleep cycle to another without waking, dogs actually wake up between each sleep/wake cycle. On average, dogs wake up twenty-three times a night—which allows them to detect intruders and alert their owners. But it can also lead to barking at every small excitement. Dogs who sleep inside seem to match our hours of sleep (80 percent of the night), while those who sleep outside spend only 70 percent of the night asleep, and dogs who sleep outside with no fence only sleep 60 percent of the night.[1]

Some researchers have found that the circadian rhythm of dogs is flexible—they can work at night as security guards or narcotics detectors without suffering from the effects of graveyard shifts like we do.[2]

But overall, there is not much consensus about how long dogs sleep each night.[3] Estimates vary from seven to thirteen hours, and some of these estimates include naps, while others do not. Adding to the uncertainty, sleep patterns change with development. Infants sleep longer and spend more time in REM sleep, which some researchers have taken to mean that REM sleep plays an important role in brain growth and development.[4] Human babies sleep for up to seventeen hours a day—and half of this time is spent in REM sleep. Our sleep does not decrease to adult levels until well into adolescence.[5] Studies from the 1960s suggest that puppies are asleep for most of the day and night, and spend 95 percent of their time in REM sleep.[6] However, there is very little data on how many hours puppies sleep and how this changes as they grow.

CAST OF CHARACTERS

Fall '18

CONGO

DUNE

AIDEN

ASHTON

Fall '19

ZAX

WESTON

YOLANDA

ZINA

ARIES

ANYA

YING

Spring '20

WISDOM

WESTLEY

YONDER

ZINDEL

ZOLA

ARTHUR

AURORA

Fall '20

SASSY

STANLEY

RAINBOW

SPARKY

Fall '21

FEARLESS

GLORIA

ETHEL

GILDA

DUNN

Brian and Vanessa with the volunteers and the class of Fall '21

Congo in the puppy park, waiting for recess

The puppy puddle of Fall '19

Spring '20—before lunch

Spring '20—after lunch

Yonder sleeping on Wisdom

Zindel being crazy

The retrieval game

The impossible task

Gilda, playing with Lion

Gloria, unsure of animal robots

Congo demonstrating infinite patience

And even more patience

Basketball is obviously an important part of puppy education at Duke. The Duke women's basketball team has "adopted" a puppy every semester. So far, every WBB puppy has graduated. Here, Coach Kara Lawson teaches a puppy the basics.

Coach Scheyer navigates chaos on the court.

What people think Puppy Kindergarten is like

What Puppy Kindergarten is actually like

Sassy and Rainbow getting ready for their morning nap

Sleep Training

Rainbow does not sleep. By that, we mean she does not sleep at night. She sleeps fine during the day. She likes her crate and happily settles down for her morning and afternoon naps. She spends time alone without complaint.

Then, at around seven P.M., Rainbow starts to get drowsy, and snoozes on and off until she goes into her crate at eleven P.M. Several hours later, we are woken up by a full-fledged creature of the night. When we come downstairs, still half asleep, to let her out, Rainbow is overjoyed. She assumes we want to come with her, that we will set off together on a grand adventure. She doesn't seem to care that her circadian rhythm is different from ours, or that as an inside dog in a family home, her sleep cycle is supposed to adjust to match the humans she lives with. Rainbow is like a bat. She doesn't just wake up at night. She comes *alive* at night.

In almost all ways, Rainbow is unlike any other puppy we have ever had. At nine weeks old, Rainbow does something we have never seen before: She climbs out of her x-pen. This is an unmitigated disaster. X-pens are the backbone of the Puppy Kindergarten. We use them for everything—feeding, testing, time-outs during rough play. Every one of the twenty-three puppies before Rainbow has respected the x-pen as an insurmountable barrier. It is only four feet high but when you are ten inches tall, this is equivalent to a child climbing onto a roof.

Rainbow scales her pen like she was born on a cliff face. When she lands as silently as a cat burglar, she looks us in the eyes. And from this moment on, we all understand that no fence will contain her.

We raise the stakes, literally. We extend the height of the fence posts by a foot and reinforce the gate. Rainbow, skinny and featherlight, uses her claws like fingers to hook into the netting and flips herself over the fence. We create complicated cardboard cliffs, complete with three-dimensional overhangs. Rainbow defies gravity. She is part squirrel and part monkey.

Even though Sassy is a week younger than Rainbow, she is much heavier, and her fluff doubles her size. Sassy is too lazy to walk from one side of the yard to the other. If it is too hot or too late or too early, she will just flop down in the mulch and wait until we carry her inside. While Rainbow performs ever more complex acrobatics, Sassy just watches, bewildered.

Rainbow increases in skill and cunning. When we finally succeed in making the fence insurmountably high, she takes a deep breath and goes underneath. It is astounding to see her pour herself through a gap the size of a lemon seed. She is a shapeshifter, her bones made of liquid mercury. Again, Sassy watches her with a look on her face that says, "Interesting. But why bother?"

Sassy, impersonating a stuffy, and our son

Sassy follows the pattern of the other twenty-two puppies we have had through the kindergarten. She wakes up a few times during the first week, at one-thirty one morning and at three a few mornings later. In both cases, she goes to the bathroom and quietly goes back to bed. Then she sleeps fairly consistently for six hours a night until she is twelve weeks old, and increases to seven hours a night by fourteen weeks old.

Stanley, Sassy's teddy bear brother, is being raised by our graduate student and is even easier. He is this semester's Mr. Perfect. Stanley requires no sleep training and very little care. He trots into his crate at night without a whimper and snores until eight in the morning. Sometimes he even rolls over and goes back to sleep.

Stanley is so perfect, he even gets along with cats.

If it were just Sassy and Stanley, we might assume that puppies from this litter have good sleeping genes. But then there is their brother Barky Sparky, who lives with our lab coordinator. Sparky is jet black and built like a soldier, muscular with perfect posture and a thousand-yard stare. If you had to pick a puppy out of a lineup to take into battle, you would choose Sparky.

And you would regret it. Because Sparky may look like a general, but personality wise, he is an absolute baby. Sparky does not like his doggy leader or going on long walks. He does not like to pee in the rain or get his feet wet. Most important, Sparky does not like anything related to sleep. Sparky does not like nap time, and he does not like his crate. He does not like the laundry room, where his crate is set up. He wakes at insane hours through the night.

Sparky voices his concerns by barking. Continuously, for a long time. Canine Companions dogs are usually fairly quiet—they don't bark often and if they do, they don't bark for long. How can Rainbow and Sassy, raised the same way in our household, have such different sleeping patterns? And why do puppies from the same litter, like Stanley, Sparky, and Sassy, sleep so differently?

There comes a time when every sleepless puppy parent has to make a choice—attachment parenting or cry it out.[7] Attachment parenting is an approach to raising human babies that promotes physical closeness and responsiveness. Sometimes attachment parenting involves co-sleeping, but it can also just refer to a general method, where if the baby cries, their parent soothes and comforts them. The parent might change their diaper or feed them, eventually settling the baby back to sleep. The philosophy behind attachment parenting is that quickly responding to a baby's distress will promote more secure attachment and less anxiety at night.

The cry it out method is also called the extinction method.[8] This approach involves putting the baby in their crib when they are around six months old and letting them settle themselves to sleep, even if this involves some fussing and crying. The philosophy behind the extinction method is that babies left to cry in a safe and secure environment will cry less over time, learn to self-soothe, and eventually stop.

After decades of research, there is still no consensus

over the effectiveness of attachment parenting (as it relates to sleep) versus the extinction method. No one we are aware of has tested these two approaches in puppies.

In the Puppy Kindergarten, we usually employ the attachment method until the puppies are twelve weeks old. Before then, if a puppy wakes up, we might let them cry for five minutes and see if they settle down. But if they sound distressed or really start to cry, we take them out to the bathroom, and walk them around for five minutes, and then bring them straight back inside to their crates. No hugging, kissing, or fussing over them—wake-up time isn't party time.

If a puppy still isn't sleeping through the night at twelve weeks old, we switch to the extinction method. Keep in mind, if you try this, your puppy's poop must be solid. Even the slightest bit of diarrhea means you are getting up in the middle of the night to let the puppy out. Also, they can't have any sign of a urinary tract infection, so look for frequent squatting without much pee, or frequent accidents inside.

Barky Sparky is sleep trained. When he is twelve weeks old, he is left to bark it out for a few nights, and after a week or so, he starts sleeping like a normal dog.

Rainbow will not be sleep trained. She has a mild UTI, followed by some diarrhea, but even after she has perfect bowel movements, she wakes up two to three times during the night. Every night. If we do not let her out, she barks every fifteen minutes. All. Night. Long. One week passes. Then two. Then three.

We try everything to get her to sleep. In desperation, we have her on the schedule of an athlete. She walks several miles in the morning and another mile in the afternoon. She does uphill wind sprints, twenty minutes at a time, chasing a ball. We take her outside at ten-thirty P.M. and play with her until she passes out on the mulch.

We become worried about her sleep deprivation. But when we look at their activity monitors, it turns out all four puppies sleep the same amount. Rainbow just naps during the day to catch up on her sleepless nights.

It's a hard lesson when you learn that your parental influence is overrated. In trying to teach Rainbow to sleep, we've just about broken ourselves.

But that is the point. It doesn't matter how recently you've had a puppy, or how many puppies you have raised, you always forget how thrilling, exhausting, and insane it is to raise a puppy. And even if you remember everything perfectly, each puppy is so different that some of what you learned with the last puppy will not work with the new puppy.

Rainbow and Sassy, reading the morning newspaper

CHAPTER 10

Memory at Work

In humans and other species, sleep and memory are largely connected. Memory is important for puppies for many reasons. Puppies need a good memory so they can remember skills and which action pairs with each word. Puppies need to remember where they are allowed to go and places that are off-limits. They need to remember what they can and cannot eat, toys they can and can't play with. But how do memories form? What is the difference between short-term and long-term memory? And what kind of memory is most important to service dogs and how would we measure it?

Fall '20 is in hybrid school. The puppies take online classes with Ashton Roberts, a puppy program manager and trainer at Canine Companions, who zooms in from Orlando. Ashton checks on their progress, gives demonstrations, and updates us with any changes in skills or puppy-raising protocol. On their assigned day, each puppy goes into the Puppy Kindergarten, where we have taken the same precautions as schools across the country to safely reopen.

Today, it is Rainbow's turn for class, and it is time to test her memory.

Technically, memory is learning that has persisted over time—information that has been stored and can be recalled.[1] But this simple definition does not encompass the range and complexity of cognitive abilities that work to form and, in some cases, forget memories. The underlying neurobiology of memory is incredibly complex.[2] Yet the process of memory formation can be described in simple terms. You experience something through your senses—sight, sound, touch, taste, or smell. Since you have sensory input from every moment in your day, to create a memory you need to encode that input into your working memory. Working memory is the conscious, active processing of incoming sensory information, for example when you try to remember a phone number. The number is only in your working memory for under thirty seconds, unless you really try to retain it. On average, your mind can hold between four and seven pieces of information at a time. After thirty seconds, these pieces of information either decay or get transferred to long-term memory.[3]

Long-term memory is the vast database of information that makes up your life experiences. You can access your long-term memories in different ways: recalling them, recognizing something in the present that reminds you of something you experienced in the past, or refreshing memories by relearning your past experiences. Long-term memory can be procedural, like playing a song on the piano or typing on a keyboard. It can also be episodic, specific to a certain event that occurred.[4]

One of the most common questions we get at the kindergarten is whether a dog remembers us if we leave for a long time. Dogs are famous for recognizing owners from long ago, at least anecdotally. Darwin was convinced his dog remembered him after he returned from a three-year voyage on the *Beagle*. Cina, a dog who we raised as a puppy for six months then gave to Vanessa's mother, remembers us when we visit Australia, even if a year or two has passed. If you feel like having a good cry, the internet is full of videos of dogs greeting their owners after they return from years of military service. There is some evidence that your dog remembers what you look and sound like even when you're absent. Research also suggests dogs can remember their mothers. In one study, several dogs were weaned from their mothers and moved into a family. Two years later, these dogs were introduced to their mother and a female dog who looked similar. The young dogs preferred to approach and greet their own mothers. The dogs also preferred to smell a cloth with their mother's scent over a cloth an unfamiliar dog had touched. However, in the same study, dogs did not remember their brothers or sisters after the same period of separation.[5]

Our ability to study long-term memory at the Puppy Kindergarten is admittedly limited. How many memories can a puppy form when they are only a few months old? Fortunately, we do have one study subject who has lived a long and interesting life: Congo.

At seven years old, Congo is entering his autumn years.

He has a few gray whiskers on his muzzle and a sprinkling of silver around his eyes. However, if anyone ever called him an old dog, he would be deeply offended. He has mentored twenty-three puppies through the kindergarten, including the pandemic class of Fall '20. He is spry for an elder; he easily walks three miles in a day, as well as doing several short bursts of cardio during play. He is lithe and lean, thanks to a careful diet, and his coat shines and his brown nose is wet.

His distinguished career saw him pass his service dog certification, not just once but twice. As a service dog, he was trained on nearly fifty skills, and some of these skills he had not been asked to perform since he assumed his role at the Puppy Kindergarten, four years ago.

When Ashton Roberts, the puppy program manager and trainer for Canine Companions, visited us just before the pandemic, we realized we had an opportunity to test long-term memory in dogs. Ashton is one of those people who has a special way with dogs. As soon as she brings Congo into the classroom, he transforms. Her voice, her posture, her gestures, transport him back to his youth. With Ashton, Congo draws himself up to his full height and comes to attention, once again a cadet in training on a sunny campus in California.

Without any practice or warm-up, Ashton asks Congo to perform the very same tasks that, long ago, he'd learned to perform to help transform someone's life. He offers a paw in greeting and politely shakes her hand. He stands up tall on his back legs and turns on the light with his front paws.

He picks up a pen that Ashton has dropped—he could

just as easily pick up a quarter or a set of keys—and holds it delicately in his mouth until she asks for it.

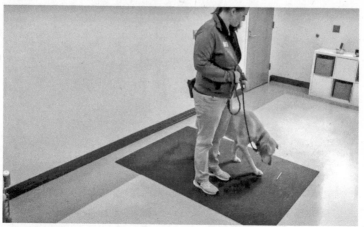

"Congo, get!"

He circles the room and closes the cupboard doors. He brings his leash to Ashton, and when she accidentally gets the leash tangled around his legs, he untangles it for her and hands his leash to her again. He maneuvers around the room expertly; if Ashton was in a wheelchair, he could thread her through the eye of a needle. He executes perfect U-turns, glides across the floor, and turns right-angled corners. Finally, when Ashton is done, he turns off the lights and, like a gentleman, opens the door for her.

Of the forty-four requests that Ashton made of Congo, there were eighteen that he had not heard since he came to live with us, four years ago. Of those eighteen, he remembered eleven the first time Ashton asked him—an impressive retention rate of 61 percent. We never imagined Congo could remember these skills so clearly after so many years

"Congo, lights!"

had passed without any practice or reminders. Suddenly, a dog remembering the Greek hero Odysseus after two decades, when no one else recognized him, does not seem as mythical after all.

Although working memory is different to long-term memory and uses different cognitive mechanisms, working memory and long-term memory do interact. Working memory also happens to be one of the cognitive measures that we have found predicts service dog success.[6]

In the Puppy Kindergarten, we primarily test working memory: a cognitive system with a limited capacity that can hold information temporarily.

To test working memory, we play a couple of simple games that build off the same principle: How well can the puppies remember where we hide a treat? In the first game, Kara hides a treat under one of two bowls. Then she waits twenty seconds before she says "Okay!" so the puppies know they can go and choose which bowl the treat is under. At twelve weeks old, Rainbow chooses the wrong bowl every time. She only has to wait twenty seconds, but her working memory is unable to retain the correct information. Though she does better on the next three trials, in the end, her score is below the average for puppies her age. By contrast, Sassy—who looks bored the whole time—scores perfectly.

Next is a more difficult memory game. The puppies must watch Kara as she hides a treat in one of two bowls again, but this time, she whips out their favorite toy, a snake made from a fire hose. Kara wiggles it around, challenging the puppy's memory by testing whether they can still remember where the treat is despite the distraction.

When it is Rainbow's turn, she is beside herself, leaping and pouncing on the snake. Rainbow does better on this test than the previous one, but she does not remember the correct bowl every week. She misses one at ten weeks, and then again at session sixteen. Overall, Rainbow performs solidly below average. Sassy, by contrast, is excellent.

In the hardest task, Kara replaces the snake with the puppy's favorite toy of all time, a green squeaky bunny. Rainbow cannot even see the green bunny without going into fits of excitement. First, Sassy looks at the bunny with complete disinterest, then she looks away, bored. When her twenty seconds are up, she wanders straight over to the correct bowl.

The different performances of Rainbow and Sassy in these games show how much memory varies from puppy to puppy. Some puppies, like Sassy, will remember information without any apparent effort, but some puppies, like Rainbow, cannot even remember something they saw twenty seconds ago.

Congo and the pandemic class of Fall '20

Memory, Sleep, and Dreaming

In some ways, it is fitting that Rainbow has the worst working memory, because of the connection between memory and sleep. Rainbow may sleep as many hours as the other puppies but it is intermittent, broken sleep. She is still not sleeping through the night, and we resign ourselves that she never will.

In people, there is evidence that when you don't get enough sleep, your working memory suffers and it is more difficult to learn.* There is little data on the relationship be-

---

* In other animals, one of the best studies on the relationship between sleep and memory was with chimpanzees. A researcher hid food under one of three cups, and the chimpanzee had to wait until they were allowed to choose—similar to the memory game we play with the pup-

tween sleep and memory in dogs, but one study looked at
the number of sleep spindles, a particular pattern of brain
waves during sleep. Sleep spindles are created by periodic
bursts of neurons through the thalamus in the lower center
of the brain, and they seem to play a part in learning and
memory consolidation. In the study, a dog learned several
commands and then had a nap. The researchers measured
the dog's brain waves as they slept, and the more sleep spindles that appeared during sleep after a learning session, the
better the dog was able to recall the commands.[8]

Whenever we bring up sleep and memories, dog owners
always ask whether their dogs dream. The short answer is
probably yes. While dogs are asleep, they have the same
types of brain waves during their sleep cycle we do while we
dream.[9] But what dogs dream about is still a mystery. The
closest researchers have come to seeing an animal's dream is
in rats. Researchers recorded the brain waves of a rat as the
rat ran through a maze. During the rat's sleep cycle, the researchers saw the same pattern of brain waves that had occurred while the rat solved the maze, suggesting that the rat
was dreaming about solving the maze.[10]

---

pies, except instead of waiting twenty seconds, the chimpanzee had
to wait either two minutes, two hours, or a whole twenty-four hours.

As you would expect, chimpanzees could remember where the
food was with more accuracy after two minutes than after two hours.
But the surprising finding was that after twenty-four hours and a good
night's sleep, the chimpanzees remembered which cup the food was
under *better* than after waiting only two hours. This led the researchers to conclude that sleep is important for memory consolidation, the
process where working memory is transferred to long-term memory.[7]

After studying many sleeping rats dream-solving the same maze, the researchers could eventually pinpoint exactly where the rat was in the dream maze, just from the pattern of brain waves. Incredibly, when the researchers sent electrical impulses into the rat's brain while they were in the dream maze, researchers could actively make the rat turn in certain directions in the dream maze.[11]

The fact that the only recorded animal dream is one where a rat was practicing solving a problem during sleep indicates that perhaps sleep does play a part in memory and learning.

Rainbow sleeps.

Memories and problem-solving abilities improve after a good night's sleep, and maybe our dreams are part of this process. Perhaps, like in rats, our dreams help consolidate lessons we have learned or reinforce solutions we have discovered. Our dreams also recount memories; people we haven't seen in a long time, a place we love but have not been able to visit. Yet it is still too early for us to define any connections between memory, sleep, and dreams in the Puppy Kindergarten. And so we will also continue to study the connection between puppies who sleep through the night and how well they perform in working memory games. But in the meantime, if you could draw conclusions with a sample size of one, Rainbow—our worst sleeper with the worst memory—would be all the proof we need.

Making Sense of It All

Rainbow's and Sassy's favorite game is to chase each other around the garden loop, a path that circles the yard, passes the children's tree house, then curves down toward the pond and back up to a sitting bench. Sassy chases Rainbow around the loop until Rainbow swan dives into a giant leaf pile underneath the tree house and Sassy crashes in after her.

Suddenly, Rainbow misses the leaf pile and accidentally slams into an old tent the children left lying in the backyard. Rainbow skids under the tent and pops up inside. She goes very still. Sassy runs around the edge of the tent, sniffing the ground, trying to figure out where Rainbow has gone.

Rainbow sticks her head out and catches sight of Sassy's bottom heading in the other direction. Rainbow withdraws back into the tent. Then she pops out in front of Sassy— who leaps a foot in the air—then disappears again. Sassy approaches the edge of the tent where Rainbow has disappeared, bewildered.

Rainbow sticks her head out again, and Sassy jumps

back, eyes wide in alarm. Rainbow disappears. Slowly, Sassy carefully wriggles the tip of her nose under the tent. All of a sudden, her tail wags frantically. She combat crawls under the tent. After a moment of silence the tent suddenly comes to life, as though a hundred genies are trying to break free. There are booms and crashes and happy barks. The tent lifts with the force of their play and floats into the air before it crashes back down.

Rainbow bursts out, Sassy close behind her. They run around the loop and instead of jumping into the leaf pile, Rainbow dives into the tent. This time, Sassy does not hesitate. She dives in after her. The tent comes alive again.

This is it—the magical moment where every cognitive ability we have so painstakingly measured is gloriously on display. Rainbow's theory of Sassy's mind, where Rainbow knew Sassy could not see her and did not know where she was—this kick-started the game. Rainbow's lack of self-control as she dives through the tent, and Sassy's more cautious investigation. Sassy's working memory telling her where Rainbow had disappeared, interacting with her understanding of the physical properties of solids like the tent, which was in turn interacting with her understanding of object permanence—that a solid object like Rainbow does not just disappear, even if you can no longer see it.

Rainbow is nearly eighteen weeks old, and Sassy is seventeen weeks old. Their brains have almost reached the end of the final period of rapid brain development. Their minds are just about equipped with everything they will need to face the world. Soon, like all the other puppies, they will graduate kindergarten and start another adventure with their new puppy parents, who will carefully train and guide them into adulthood.

To Rainbow and Sassy, this backyard, Congo, and our family will soon be just memories. Which ones will filter through to their long-term memories? Will they remember the nights we cradled them to sleep, Congo's wet nose in the morning, or the walks and adventures we had together? Perhaps, years from now, they will remember their new game, discovered as every cognitive skill converged into a moment of joy.

CHAPTER 11

Our Takeaways

What We Have Learned from Our Puppy Kindergarten

As of May 2024, we have followed the development of 101 puppies, whom we tested every two weeks, while they were between eight and twenty weeks old (during their final period of rapid brain development). Of this sample, we raised 52 puppies in twelve different classes at the Puppy Kindergarten. The rest of these puppies were raised in family homes by Canine Companions volunteers. We also tested another 221 Canine Companions puppies a single time as well as 37 wolf puppies. We also tested hundreds of adult pet dogs, military working dogs, and service dogs. And finally, we have data from almost fifty thousand pet dogs that were tested by their owners as part of the citizen science project Dognition. All this work has led to a clearer understanding of how dogs develop and how we can raise a great dog.

Our most important discovery at the Puppy Kindergarten is that the puppy brain does not function like a lightbulb that just switches on. Nor is it a gradual illumination, where a puppy's intelligence slowly matures at an even pace. We found no evidence for general intelligence, otherwise known as IQ. Instead, a puppy's brain is like a symphony of lights— each cognitive ability switches on at different times. Some skills appear early and almost immediately shine brightly while others take longer to show a first glimmer and reach their full potential.

When we started our project, we knew that a puppy's motor cortex is the most developed part of their brain at birth. At two weeks, puppies open their eyes and the visual cortex rapidly matures. By around six weeks their brains can process movement, sight, smell, and sound. We also knew that by twenty weeks old, the puppy brain has finished its last rapid period of brain growth. But before our project, the changes that occur in the puppy mind from eight to eighteen weeks old remained mostly a mystery. For the first time, we can pinpoint when certain important cognitive skills come online—cognitive skills implicated in training success of service and military dogs and that are important for your puppy too.

EIGHT WEEKS

We found that at eight weeks, when weaning is already finished, a puppy's eyes begin to have adultlike acuity and pass both our visual and odor detection tests. Puppies begin to remember where they have seen things disappear. If you hide a treat, even after a delay, puppies remember where it went. In the more difficult tests, where they are distracted by a squeaky toy, they can still recall where the treat went. At eight weeks old, memory has emerged.

Consistent with our previous work comparing wolf and dog puppies, we found that puppies already show basic memory and communication skills when they first arrive at eight weeks of age. This is when puppies begin interpreting basic human gestures: If you hide a treat and point to it, your eight-week-old puppy can probably find it. It is not just because they have already seen you point. They are just as skilled using other gestures they have never seen before, like if you place a physical marker next to the hiding place.

Aiden and Ashton at eight weeks old

TEN WEEKS

Just two weeks later, puppies have mastered the use of basic human gestures and demonstrate an adultlike level of comprehension. The first cognitive skill to clearly emerge and be mastered by puppies is not self-control or an understanding of the physical world. Instead, it is a social skill: the cooperative-communicative ability to read human gestures—also critical in our early development. It is a vital cognitive skill that puppies will use for the rest of their lives to socially learn from people. The failure of wolf puppies to use similar gestures supports the idea that this cooperative-communicative ability emerges early because of domestication.

Left to right: Westley (ten weeks), Zola (nine weeks), and Yonder (nine weeks)

In the short time it takes puppies to master cooperative-communication, other abilities begin to emerge. In particular, our puppies begin to solve the basic self-control problem that requires them to take a detour around the front of the cylinder at ten weeks of age. It is the first evidence that a puppy can override the urge to make a tempting but incorrect choice in favor of one that takes more time to pay off but leads to success. Early emerging self-control is a vital ingredient to many forms of more complex problem-solving.

THIRTEEN WEEKS

Until puppies are thirteen weeks old, if they come across the impossible task, they tend to concentrate on solving the problem on their own. As their self-control continues to develop, puppies begin to show a more socially complex strategy. After failing to open the snap-lock container holding a treat, puppies start to make eye contact. This is the first emergence of behavior that allows adult dogs to ask a human for help—a skill a service dog, and many pet dogs, will rely on as they cooperate and communicate with their owners. It is also a behavior we do not observe in wolf puppies and has probably been shaped by domestication.

At around thirteen weeks old, puppies start to show an understanding of physics, or causality. If you hide a bowl with food under one of two towels, puppies start to understand that they should search under the towel with a bowl shape, instead of the towel that lies flat on the floor. This is a sign that puppies are beginning to understand properties of the physical world—in this case that solid objects cannot pass through one another. As puppies continue to develop, they also gain a simple understanding of other causal properties such as gravity and connectivity (that objects act together when connected—like when they are connected to you with a leash). What puppies understand about the physical world can limit or facilitate the types of problems they can solve and be trained to solve.

Class of Fall '19 at almost thirteen weeks old

FOURTEEN WEEKS

The final cognitive skill to emerge during the puppies' time with us is the ability to reverse what they have learned. This is a more advanced form of self-control than the first part of the cylinder task, when they simply have to take a detour. In the reversal learning game, a choice that was correct becomes incorrect—in our case, we close one end of the cylinder. Over several trials, puppies have to stop themselves going to the side that used to be open, and reverse their learning to make the opposite choice. This can even be hard for young children, and it signals that a puppy is ready to start learning to solve more complex problems.

OVERALL PATTERNS OF EMERGENCE AND MASTERY

When we look at the performance of all 101 puppies, an overall pattern emerges. There is a first stage of develop-

Fall '19 at seventeen weeks old

ment between eight and ten weeks of age. Puppies at this age are ready to use their eyes and nose to solve a range of simple problems. Puppies also quickly develop the use of basic human gestures. Remarkably, by ten weeks they show adultlike skill with the same basic gestures. At ten weeks, self-control begins to emerge. Between twelve and fourteen weeks, puppies begin making eye contact, are able to understand basic causal problems, and learn how to reverse what they have learned. They also develop adultlike mastery in the simpler memory and self-control problems.

Of course, there are exceptions to these generalizations. There are genius puppies who crush all the tests. Wisdom was one of these. She achieved mastery of almost all the cognitive tests before she was even thirteen weeks old.

Then there are puppies who are genius communicators but struggle with physics. There are puppies who get better at certain games over time, and some who get worse. Then there are puppies where we have no idea what is happening inside their minds for the whole time they are with us. But on average, the figure below shows the general pattern of cognitive development.

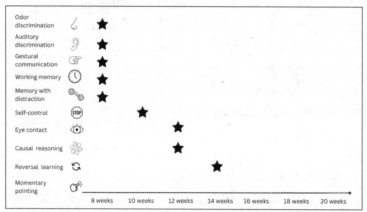

Stars represent the age of puppies when "first emergence" was observed for each of ten cognitive tasks associated with training outcomes in adults. "First emergence" is defined as the age at which 50 percent of puppies perform above chance. Only momentary pointing does not first emerge in the age range we studied.

BREEDING AND REARING

In addition to identifying cognitive skills that predict a dog's training success in the real world, our team and others have collected data that indicates self-control and cooperative communication are highly heritable.* Other abilities like

---

* Remember that heritability is the proportion of variation in a certain trait that can be attributed to genetic variation as opposed to environmental causes. We found cooperation communication and self-control to be over 40 percent heritable. Meaning, genetic differences between dogs explain 40 percent of their cognitive differences on tests like self-control or their understanding of human gestures.

memory and a dog's capacity to navigate their physical environment are also heritable but genetic differences explain less about the individual differences in performance. These results illustrate how cognitive individuality in dogs is not just a result of how we raise them. Genetics influence who our dogs will become. Continued developments in genetic analysis will help further identify the specific genes involved in building a dog's brain—those that are likely involved in producing some of the differences we see across individuals.

Studies of heritability also reveal that the majority of cognitive differences across individuals are the product of other factors beyond genetics.[1] A significant part of this variation must be influenced by experience. That is why we designed our program to give our puppies two different types of rearing experiences. We know that puppies deprived of socialization during the critical window of six to eighteen weeks old can have big problems with their temperament and cognition. However, no one knew whether a high level of socialization could increase the likelihood that a puppy would develop the cognition and behavior that lead to training success. If socialization is a good thing, is super socialization better?

Our initial results suggest that—happily for everyone at home—a puppy does not need to be a Duke puppy to become a great dog. Our home-reared puppies performed just as well as the kindergarten puppies who had extensive socialization and exposure on campus. This seems to indicate that there is a threshold for socialization of puppies—including yours. As long as your puppy interacts with other dogs and people during their socialization window, they should be able to reach their potential.

In the future, we will follow our 101 puppies as they mature and see if their performances can help us predict who they grow up to be. We are betting that playing games that test what they think and feel as puppies will tell us a lot about how they will think and feel as adults. Of course, these discoveries are new and science can be slow, and we have yet to determine long-term possibilities, whether we can deploy our games into the real world and get more dogs helping more people. But we have reason to be optimistic. The individuality that makes us fall in love with our dogs is the same individuality that can help us better identify dogs with the most potential to succeed.

What These Lessons Mean for You

At 101 puppies, we want to share the main lessons we learned at the Puppy Kindergarten that we think might be most useful for you.

1. MAKE EYE CONTACT

Making eye contact is one of the earliest and easiest ways you can bond with your puppy. Oxytocin is a hormone involved in bonding between parents and their children. When your puppy stares at you, and you stare back, evidence suggests it creates an oxytocin loop, making you both feel loving and loved. We have shown that playing a game that involves eye contact for a few minutes every two weeks increases the amount of eye contact a puppy makes. If your puppy is a bit shy about making eye contact, try getting their attention with a small treat and holding it under your eye. See how long they can hold eye contact and if it improves over time. Sometimes you might notice your puppy sitting and staring at you for no reason. They are just trying to make eye contact with you. It's their way of hugging you with their eyes.

Aiden making eye contact

2. PLAYTIME

The traditional method of parenting puppies is pretty hands off. Puppy parents might take their puppy on a walk or play fetch with them, but puppy-parenting culture does not involve a lot of active play. However, just like young children can benefit from active engagement with their parents,[2] we suspect that the same is true for puppies. Throughout the book, we try to explain the experimental methods for our games clearly enough that you can play them at home with your puppy. It doesn't have to be precise—it should be fun for both of you. You can also find these games on Dognition .com and on our free online course "Dog Emotion and Cognition." You might be surprised at how they see the world and you might even uncover their special genius.

3. MIND READERS

At only eight weeks old, your puppy can understand your cooperative-communicative intentions better than any other species on earth, including your great ape cousins. Their brains are still developing, but at eight weeks old, puppies begin to understand your basic gestures. You can use this to play hide-and-seek with a toy, or to point out the names of different objects while training. There are going to be lots of times you want to show your puppy something, for example different objects around the house, people who are important to you, or something they have missed. When it comes to your puppy, actions speak louder than words.

4. PUPPIES ARE NOT EINSTEIN

Puppies are not very good at physics. Examples include leashes getting tangled, balls rolling down hills, navigating around doors, what happens when tails come into contact with fragile items—the list goes on. Sometimes puppies improve as they get older, but in general, dogs are not very good at physics so you will probably have to do some thinking for them. Puppies are also not very good at understanding what is dangerous. This means you should prepare and maintain your house as though a toddler is living with you. No sharp objects, don't leave food or cables lying around, and if your puppy starts roaming around the house, make sure they are supervised.

5. SELF-CONTROL

We've seen that some cognitive skills emerge early, and others take time. Self-control is one of those skills that might appear at around ten to fourteen weeks of age, but it increases gradually. This means that puppies are impulsive and reckless with short attention spans. When this combines with their limited understanding of physics, you will occasionally have puppy disasters. The number one lesson is to be patient. Your self-control is fully developed, and theirs will come with time.

6. THEY WILL EVENTUALLY SLEEP THROUGH THE NIGHT, WE PROMISE

Sleep is the biggest challenge we deal with in the kindergarten. We define sleeping through the night as puppies spending seven consecutive hours in their crate at night. Puppies are on a much shorter sleep cycle than people. Puppies wake up more frequently and when they wake up, they are more awake than people during these sleep cycles.

It is hard to be a good parent when you are sleep deprived. We have not solved this problem completely, but we have found some things that help.

First, if your puppy is frequently waking up at night to go to the bathroom, make sure you check with your vet that your puppy doesn't have a parasite, like giardia, or a stomach issue that will usually present as diarrhea. Also check for a urinary tract infection, which usually presents as a puppy frequently squatting or leg lifting but only expelling small amounts of liquid.

We have found that most puppies cry at night for a reason—usually they need to go to the bathroom. This is where the fifteen-minute bathroom break just before bedtime is essential. A rookie mistake is to come back inside after they have only peed once. Walk them around, make sure they pee and ideally poop. And remember dogs often poop in multiple bouts so the full fifteen minutes is critical for their sleep and yours. If it is late at night and they are so sleepy they just flop on the ground, carry them some distance from the house. They will usually get up to walk back to their warm bed, and hopefully they'll go to the bathroom on the way.

If you are sure your dog's restlessness is not a bathroom issue, you can try different things. Put them in a different room to yours at night. Let them cry it out for a few nights. Ask other family members to help you.

Whatever you choose to do, we have some hopeful statistics. By the time they leave the kindergarten at eighteen weeks old, every single one of our puppies was sleeping through the night—even Rainbow. It will get better, we promise.

7. SOMETIMES THEY FORGET, BUT THEY WILL REMEMBER WHAT'S IMPORTANT

Puppies can forget things fast. As in twenty seconds later they've completely forgotten whatever they have seen or what you have told them. They are not being disobedient— they literally cannot remember. Their memory does improve over time, and they do have long-term memories. As we saw, Congo could remember skills he had not used or practiced in years, and it seems that dogs can remember people they have not seen in a long time, especially someone who is special to them.

8. SOCIALIZE THEM EARLY AND OFTEN

The main socialization window in puppies adopted into a human family is from eight to eighteen weeks old. During this time, make sure they see as many people and places as possible—as long as you can ensure that these experiences are positive. To grow up with confidence, puppies should meet people from diverse backgrounds, of different ages, and in different situations. They should also meet and play with lots of different dogs—as long as you are sure that the dogs are up to date on their vaccinations (you can be more relaxed once your puppy is up to date on their vaccinations at around sixteen weeks old). When you meet other dogs, make sure these dogs are friendly and tolerant of puppies and do not show any signs of illness. Make sure your puppy has positive experiences as you introduce them to noises like traffic, the vacuum cleaner, and eventually extremely loud sounds like fireworks (from a distance at first or while

eating). Make sure you know the signs that your puppy is relaxed and comfortable, versus overwhelmed and stressed (there are many good articles and videos on this), and if your puppy starts to look uncomfortable, remove them from the situation and try again with less intensity before you build it back up. During this eight- to eighteen-weeks window, you should try to introduce your puppy to the experiences you think they will encounter as part of your family.

9. ALONE TIME

Just as important as learning to be with others is learning to be alone. When people went back to work after the pandemic, we heard many new puppy parents struggled with their puppy's separation anxiety. At the Puppy Kindergarten, we make sure our puppies spend time alone resting in their crates. It is important for them to learn to be by themselves and relax in a quiet place.

10. TAKE THEM FOR WALKS

The flip side of this is that puppy parents who work must know that they can't just leave their puppy (or a dog) alone in a crate for eight hours while they work. Someone has to come back to let them go to the bathroom, and even then, puppies have a lot of energy. So we take them for walks. So many walks. We've found that most puppy behavior problems can be solved with a walk. We've found walking particularly helpful when they are teething, full of puppy energy, potty training, or just generally being crazy. Our puppies average four walks a day. These walks are usually less than a mile, depending on the puppy. We know that this is difficult for puppy parents who work in an office, and some puppies need more exercise than others. The solution usually involves reaching out to friends or neighbors. Maybe a responsible teenager in your neighborhood would like to earn some money after school. Or maybe you have a friend with a dog you can swap walks with. Doggy daycare can be a great option for dogs whose parents work long hours (although you will want to wait until your puppy is fully vaccinated and your vet gives the okay). All puppies need to be walked at least once a day in all kinds of weather. And we've found that no matter how awful it looks outside, both the puppy and their walker always feel better once they are back.

11. EACH PUPPY IS AN INDIVIDUAL

Despite everything we have learned about puppies and dogs in general, our biggest lesson is that there is always a puppy who doesn't conform. Who is a genius in this test or utterly clueless in that one. We have had puppies who broke equipment built for military dogs and puppies who solved the impossible task. We have had puppies who reversed someone's dog phobia, and puppies who tried to nurse from Congo's elbow. Based on our 101 puppies, the one thing we can tell you for certain is that the relationship you have with your puppy is unique. And even if this uniqueness sometimes presents in the most annoying way, it will probably become what you love most about them.

Even though these puppies look the same, they are all individuals: Neely as a seal, Madeline as a mermaid, Nancy as a narwhal, and Maestro as a lobster (Fall '22).

12. YOU ARE ENOUGH

You and your family, whatever that looks like, are enough to give your puppy the perfect head start and raise a great dog. With all our experience, our hundred undergraduates, our world-class veterinarians, and our significant funding, we could not do better. Puppies raised in our kindergarten were no more successful, trainable, or endearing than puppies raised in a loving family home. You are everything you need to raise a great dog.

Polar, Spring '23

CHAPTER 12

Aging in Place

Inside the puppy park, the class of Fall '21 is out for recess. Everyone is happy to see the puppies on campus again. Congo arrives, having seen the children off to school and cleaned up the mess of Cheerios they left on the floor. He is getting older. He has a distinguished grizzle around the eyes and muzzle, and some mornings he is a little stiff.

The volunteers smile under their masks and indulgently separate Fearless and Dunn. Fearless, despite being a week younger than Dunn, is one pound heavier and uses this extra pound to gain the authority he feels he needs to dominate the puppy park. Dunn, our only yellow puppy this semester, has seen Fearless sleeping with his head in his water bowl and knows that Fearless the boss puppy is a ridiculous concept.

The squabbling between Dunn and Fearless is just a show. If there is a boss puppy of this class, it is Ethel. At ten weeks old, she is the smallest but already an aging empress. She takes forever to eat her food, snores loudly in her sleep, and demands constant attention from everyone.

Gilda and Gloria are the youngest. Collectively known as Glorilda, they are the class sweethearts. They follow the rules, don't make a fuss, and everyone is sure they will make it.

Fall '21, left to right: Ethel, Gilda, Dunn, Gloria, Fearless

Congo pauses at the top of the hill and gives the puppies a chance to see him, which they do. They rush over with their bottoms wagging so hard they can barely walk in a straight line, stepping on one another in desperation to greet him. Congo waits until they are lined up somewhat in order before he bows his head and gives each one a kiss, right on their little wet noses.

Then he walks inside without a fuss, goes into his office, and takes a nap.

One of the questions every puppy parent will have to answer eventually is how to take care of an aging dog. Perhaps you have an older dog like Congo, who is part of the family your puppy comes home to. Or perhaps you adopt an older dog. Even if you and your puppy have just met, one day they will get old.

Physical signs of decline are easy to identify. Like Congo, dogs can get stiff in the morning. Their pace gets slower, and they don't walk as far. Veterinary medicine has made great progress in caring for their bodies, but we know much less about how their minds decline and how to take care of them.

As dogs age, they can show symptoms similar to Alzheimer's—they might become disorientated on walks, bump into walls and doors, and sleep during the day but become restless at night.[1] Some types of cognitive decline can be managed, but others can be dangerous, like wandering off and falling down the stairs.

What should we expect as Congo enters his twilight years? And what exactly *are* his twilight years? We know that big dogs like Congo don't live as long as small dogs—for example, the average lifespan of a Saint Bernard is eight years, while it is thirteen years for a Yorkshire terrier. But does the Saint Bernard go through cognitive decline when they are three quarters of the way through their lifespan—at six years old? And does the Yorkshire terrier go through cognitive decline three quarters of the way through their lifespan, at nine years old?

One hypothesis is that the brain of a dog ages at the same pace as their body. According to this hypothesis, a large dog like Congo has a body that ages much faster than a smaller dog's and would also show faster cognitive decline.

The alternative hypothesis is that Congo's brain does not age at the same pace as his body and that cognitive decline is independent from physical decline.

Using Dognition data, Evan worked with Marina Watowich to test how the cognitive abilities of dogs from different breeds decline with age. When she looked at data from four thousand dogs from sixty-six breeds in the Dognition database, her analysis showed that while large dog bodies begin to age earlier, their cognition does not.[2]

Basically, all dogs, regardless of how big or small they are, start to show cognitive decline at around the same age— ten years old. This means that once smaller dogs are older than ten, even though they may be as sprightly as ever, you should watch for the effects of cognitive decline and figure out how to manage them. Often, we can make small changes to our homes to accommodate aging dogs, just as we would aging people. This might mean blocking off the stairs or making sure doors are closed so they don't wander off into the street. If you notice any odd behavior, talk to your vet.

At ten years old, Congo is in excellent physical condition. His regime has always involved a careful diet, exercise, good dental hygiene, and regular veterinary checkups. And although we know we must watch for slow signs of cognitive decline, his mind is in perfect health. He still helps with testing and figuring out new research protocols. With the puppies, he is still patient, loving, and dignified.

But in the summer of 2023, we get devastating news after a visit to the vet, and we realize you are never really prepared. Congo has cancer and it is spreading. And just like that, we are finally where no puppy parent ever wants to be—but where we all end up eventually.

Our family gathers in the living room. One of the children moves their beanbag next to Congo's bed and snuggles him until he falls asleep. The other one comes to check on him periodically, kissing his nose and stroking his ears.

We don't know why, but we thought we would have more time. Part of it is that life without Congo is incomprehensible. He has always been there. The children cannot remember life without him.

We check on him before we go to bed. We wipe his eyes and his nose; they are always running these days.

"Good night, Congo," we say. He looks at us and he is still there. He is still Congo.

If ever there is a moment when we need him to understand us, it is right now. This is the final test. The last game we will ever play.

"We love you."

He blinks, and we know he knows.

He blinks again, and we know he loves us too.
In the morning he is gone.

The downside of having a great dog is that one day you will
lose them, and the grief will knock you sideways. Part of it is
that for a decade they bear witness to every triumph and
tragedy, as well as long stretches of ordinary life.

But this is also their gift. They teach us how to lose
someone we love. And that no one we love is ever truly gone.

Congo will not see the next class of puppies line up to
greet him. He will not be there to teach them how to play
without fighting or how to politely make new friends. But

---

because Congo taught us too, we know how to do it. We remember his inexhaustible patience, his calm consistency, and his kindness. We remember that a life caring for others is a life well lived.

We are proud of our graduates who are placed. They have chosen to serve with the highest honors. Only a few make this choice, and even fewer are able to. The testing is rigorous, the task unending.

To succeed, they must have all the traits we are able to measure—and a few more that we can't. They must find satisfaction in a task perfectly executed, every time. They must see purpose in being someone to lean on, a shoulder to cry on, and the one who cheers on each small victory. They must be content with their reward—the infinite gratitude of someone who loves them.

But then, we are proud of all our puppies. They are with us for such a short time but each one has lit a thousand smiles. At the Puppy Kindergarten, we are supposed to be teaching them, but it is the puppies who teach us. They remind us how fun it is to be silly. To live each moment mindfully. And that all of us can make a difference in the small corner of the world where we live.

If you are reading this book because you have, or may soon have, a puppy of your own, may you be blessed with a puppy who sleeps through the night. May they never mess in the house, especially where you might step in it with bare feet. May they only chew designated toys and never anything expensive or irreplaceable. May they learn their skills in a day and never embarrass you in front of strangers.

But if your puppy, like all of us, is less than perfect, then may you settle for some simpler gifts.

Love without expectations.

Acceptance without judgment.

Joy with each greeting.

A light to bear witness, and to brighten your days.

And may you return these gifts in equal measure, to your puppy and all those you love.

EPILOGUE

In a pretty house in a leafy suburb in North Carolina, a dog waits for someone to come home. He is tall, with long legs, dark blond fur, and brown eyes.

You would never recognize him. Aries, the puppy who sulked in everyone's lap for a week, wouldn't walk on a leash, and didn't want to play with the other puppies. The puppy from Fall '19 that no one thought would make it, who was judged least likely to ever graduate as a service dog.

Wonders never cease.

Years ago, Christine worked at a nonprofit in Uganda that helped orphan children. While there, she met Makisa, a young girl who had severe brain damage after a bout of cerebral malaria as a baby. Christine used art therapy to help children through trauma, and she noticed that Makisa was always near and didn't want Christine out of her sight. After falling in love with Makisa's powerful spirit, Christine decided to adopt her. Together, they settled in Charlotte. Christine got married and had three other children, and Makisa finally had the family she always wanted.

Makisa went on to flourish at school and has many friends; she is content, safe, and loved, as all children deserve to be. But there were just a few things that worried Christine. Makisa is very shy. It is good for her to meet new people, but Makisa often found the initial interaction difficult. She does not like people staring at her. In their happy home, things can get loud, and Makisa craved quiet space in her room. But there are certain tasks Makisa needs help with and Christine worried about leaving Makisa alone.

Now when Makisa meets new people, Aries steps in. Handsome and calm, he attracts all the attention. Makisa quietly watches, proud that people love her new friend. Instead of staring at her, people smile and ask about Aries. He makes interactions with new people positive and helps Makisa make new friends.

At home, it turns out that Aries hasn't changed much. He always wanted to be away from the noise and drama, sitting quietly in someone's lap. His favorite place in the house is Makisa's room. When they go in together and close the door, Christine is no longer worried. Makisa is not alone, and if she needs anything, Aries is there to help her.

At the sound of the school bus Aries runs to the door. He trembles with excitement, but he has finally learned what Congo tried to teach him so long ago—service dogs don't jump.

The door opens and there is a soft voice.

"Aries?"

He stands at the door, ears up and tail wagging with the best of life's emotions: joy, gratitude, love. The soft voice seems thankful for these gifts at the end of her day.

"Aries."

Aries and Makisa

As a child, it is hard to be different. It is even harder when your differences can be seen from a distance, when it is immediately obvious that you are not like everyone else. This is one of the gifts that these dogs bring. Our puppies who graduated as service dogs and are now a part of the elite cadre of dogs to which Congo once belonged, all have one trait in common—they can charm a smile from the gloomiest stranger.

Thanks to their new companions, these children have someone to help them be brave in hospitals during painful procedures. They have someone to talk to when they can't sleep, and someone to hug when they wake up from bad dreams.

When these children go out into the world, their companions are a point of conversation, something meaningful to talk about. Their love of dogs is something they can share. And for a child who feels their differences keenly, finding something in common with others is a welcome respite.

Three of our puppies are currently placed with children:

Arthur, our Labbiest Lab, is placed with Carter, who has autism.

Aurora, his sister, is placed with Izzy, who has brain cancer.

Arthur, our Labbiest Lab, from Spring '20, with Carter

Aurora, Arthur's sister, with Izzy

Wisdom with Easton

And Wisdom, our valedictorian, is placed with Easton, who has spina bifida.

Our other puppies are placed with people of all ages. Ashton, the front man of Fall '18, is paired with Nancy. He is still as calm and unruffled as ever. A perfect gentleman, he opens doors for Nancy, who is in a wheelchair. He carries her handbag and finds her keys if she misplaces them. He is

always ready when she needs him and never makes her wait. He is patient and kind.

Little Yonder, the smallest puppy from Spring '20, finds her home with Matthew, an instructional designer with multiple sclerosis. Wisdom's brother Westley is placed as a facility dog.

And just to throw us a curveball, Barky Sparky also makes it. He is paired with Erin, a junior at Louisiana State University. After missing out on being on Duke's campus during the pandemic, Sparky gets to go to college after all.

Even after raising dozens of puppies, it is impossible to know who is going to make it. There are puppies like Wisdom, whose success surprises no one. But there are also puppies who are equally gifted, who did not make it but continue to lead enriching, joyful lives. Then there are the puppies like Aries and Sparky, whose graduation leaves us bewildered. Without our scientific approach to understanding these puppies, we would only be left with intuition: We would always just be guessing.

Our collection of cognitive tests will probably never be finished. They will continue to evolve, as they have before, and will need to be refined, adjusted, and rethought as we learn more about these incredible dogs. After all, it was designed to complement, not replace, the incredible trainers, veterinarians, and volunteers who make these dogs possible.

Once you meet even one person whose life has been changed by a dog, you will know, as we do, that all this work is worth it.

ACKNOWLEDGMENTS

T hank you for choosing our book and spending time
to learn about our work at the Puppy Kindergarten.
We hope what we shared will help you and your
puppy. We have done our best to refer to our key discoveries
and those of our colleagues in the text, notes, and refer-
ences. The scientific method is the most powerful system
for generating knowledge on our winding path to the truth.
While the literature we have cited is not meant to be comprehensive, we have done our best to cite papers that can
lead you to our original data and in-depth review papers that
point to where there is scientific debate on issues we have
covered.

It feels like a different world to the one in which we
began writing this book, five years ago. The one thing that
hasn't changed is the cast of people we have come to depend
on. We could not have accomplished anything in these pages

without you. Every triumph was a massive group effort, a product of hours of brainstorming, failures, restarts, and "Puppy, look!"

Thank you to our local service dog partners, Ears Eyes Nose and Paws, including Deb Cunningham and Maria Ikleberry. Time and space did not allow us to sing your praises but your work and your service dogs have moved us deeply and are instrumental to our research. We are so grateful for how many puppies you have dropped off for testing—a good number of our 101 puppies are yours.

Thank you to the incredible Margaret Gruen. We would never have even had the first puppy without you. Thank you for all the late-night video calls, teaching us how to bandage a splint over Zoom, letting us flood your phone with poop photos and desperate texts of *Is this normal???*

Thank you to Kara Moore, our tireless research coordinator, who gently and patiently coaxed dozens of puppies across the testing mat in thousands of trials. Thank you to the brilliant Evan MacLean, who helped us make all our initial discoveries, from the Congo Basin to K2, and became a geneticist, hormone specialist, and big data researcher along the way.

Thank you to Emily Bray, one of our first undergraduate stars. It has been so much fun to see the remarkable researcher you have become. Thank you to our amazing grad students, undergrads, and lab coordinators who have gone on or will go on to great careers: Alex Rosati, Jingzhi Tan, Chris Krupenye, Aleah Bowie, Wen Zhou, Morgan Ferrans, Hannah Salomons, Gabi Venable, Rachna Reddy, Leveda

Chang, Kerri Rodriguez, Ben Allen, Kyle Smith, James Brooks, Madison Moore, Margaret Bunzey, Sam Kefer, Jessica Shoemaker, Lizzy Glazer, Emily Sandberg, Harriet Caplin, Anya Parks, Leah Ramsaran, Sam Lee, Julianna Turner, Jordan Sokoloff, Colin Kelly, Elizabeth Wise, Gabby Bunnell, Emma Nelson, and Candler Cusato.

To our amazing friend and veterinarian, Erin Tom: Thank you for putting up with us leaving poop bags with diarrhea on your car and for dropping meds on our doorstep at 10 p.m. And thank you to the rest of the team at Colony Park Animal Hospital—you guys are the best.

Thank you to everyone who contributed puppy photos: Morgan Ferrans, Madison Moore, Jennifer Bidner, and all the puppy volunteers. Most of the photos in this book are yours and we are so grateful.

Jared Lazarus, you are the best puppy photographer of all time. Thank you and the whole team at Duke Media and Communications for the crazy class photo sessions, the graduation photos, and documenting the puppies' journey with such love and dedication.

Thank you to Mike Tomasello: You are the best friend and mentor a clueless nineteen-year-old undergrad could hope for. To Sarah Gaither and Jenna McHenry: Thank you for being awesome colleagues. Thank you to Steve and Susan Nowicki, Steve Churchill, and Rich Kay, for letting the puppies take over the sub-basement.

Thank you to Valerie Ashby, Dan Kiehart, and Susan Alberts, who took seriously the crazy idea of raising puppies on campus and made possible the Duke Puppy Kindergarten.

Our research is funded in part by grants from the Office of Naval Research (N00014-16-12682), the Eunice Kennedy Shriver National Institute of Child Health and Human Development (NIH-1RO1HD097732), and the Canine Health Foundation (Grant-#02700). Big thanks to our program officers, Lalya Esposito and Joong Kim, for believing in our work from the start. Thank you to Anna Hampton and the IACUC team for helping us provide exceptional care in our on-campus puppy kindergarten. Thank you Joe Gonzales and Debbie LoBiondo, who entertained the crazy idea of having a puppy in every Duke dorm.

Thank you to our friends Herman, Janice, Alex, and Clara, for being our pod family during the pandemic and putting up with puppies in the living room during online school. Thank you to Lisa Jones, for always figuring out how to get us out of whatever mess we are in and being such an amazing friend while doing it.

Thank you to over five hundred Duke Puppy Kindergarten volunteers, who shower the puppies with love; the administrators, and the broader Duke community who has helped turn Duke's campus into the puppies' home.

To Brenda Kennedy, Ashton Roberts, Martha Johnson, Laura Douglas, Theadora Block, and Jeanine Konopelski of Canine Companions: Thank you for putting up with our crazy scheduling, our incessant request for sex-age balance, and just generally being awesome. You guys are the best and we love you.

Finally, to all the great dogs out there who are changing people's lives. Thank you for your service.

If you would like to donate to Canine Companions, visit canine.org/donate.

If you would like to be a puppy raiser, visit canine.org /raiseapuppy.

If you live in North Carolina and would like to support a local organization that we have worked closely with through-out the years, visit Ears Eyes Nose and Paws at eenp.org /donate.

If you would like to learn more and support our research, visit dukepuppykindergarten.org.

Supplies for New Puppy Parents

PUPPY CARE

- **Crate.** We use large metal crates that you can see into from all sides and have a removable plastic tray at the bottom, in case there are messes. Dorm rooms are small so the crates can be stored up and folded when the puppies are not there.
- **X-pen.** An x-pen is a small plastic or metal fence that can be quickly put up and taken down if the puppies go somewhere and need to be contained (e.g., in a common space in a dorm or outside during an event). These are essential when the puppy needs to be left alone for a minute (e.g., if their volunteer needs to go to the bathroom). We use these in places where the puppies are unsupervised.
- **Toothbrush and toothpaste.** Puppy teeth are brushed every night with toothbrush and toothpaste meant especially for dogs.

- **Ear wipes.** Puppy ears are cleaned every week. We have found that puppies like the wipes better than the liquid.
- **Grooming brush.** We use rubber soft brushes on puppies every day.
- **Nail clippers.** Puppies have their nails clipped (very carefully) each week.
- **Puppy shampoo, wet and dry.** Puppy shampoo is different from human shampoo. You can get the wet versions that require water—good for when they get really muddy—and the dry version that you rub on if they stuck their nose in something smelly.
- **Leash (four feet), collar, and a doggy leader.** These are used every time the puppy goes outside. When the puppy is young, a four-foot leash is helpful so the puppy doesn't get tangled. A six-foot leash can be introduced later on.
- **Poop bags and dispenser.** To be attached to the leash. No one can remember to take a poop bag every single time.
- **Diaper genie.** These are handy for inside messes in apartments, because the poop and pee (and all the associated paper towels) should not go into a normal bin.
- **Odor cleaner.** This is essential for carpet and floor messes.
- **Recycled paper towels.** We go through enough paper towels to deforest the Amazon. Buying recycled is the least we can do.
- **Disinfectant spray and disinfectant wipes.** We use Rescue because it kills bacteria and pathogens like parvovirus on contact. It is also scent free.
- **Laundry basket of old towels by the door.** Sometimes messes can go beyond a disinfectant wipe.

- **Go bag.** The go bag is for whenever the puppy goes somewhere, whether to the dorms or on an outing. The go bag has a water dispenser, disinfectant wipes, a chew toy, treat bag (with kibble), and, of course, poop bags.

EATING AND DRINKING

- **Puppy food.** A well-balanced, age- and size-appropriate diet (e.g., one formulated for puppies and a larger breed dog, in the case of our puppies). We use Eukanuba but any high quality kibble will do. We do NOT recommend an all-meat diet (see Appendix III); it is not necessary for dogs and it sucks for the planet. You can also feed your puppy table scraps, but talk to your vet first.
- **Storage.** If the bin is inside, we use a plastic bin, but at home our puppy food is outside, so we use a metal garbage can, which is rodent and cockroach proof.
- **Water bowls.** We have water bowls everywhere the puppies spend time, because dehydration is especially dangerous for puppies.
- **Food bowl.** Puppies are fed inside their crates, so we use a small bowl that is washed after each use. Don't just let it sit there—it gets gross.
- **Treat bag.** The bag can be filled with kibble.

PUPPY MEDICATION KIT

- **Styptic powder.** To quickly stop bleeding, especially on puppy toenails.
- **Gauze.** To bandage cuts.

- **Diphenhydramine HCl** (the active ingredient in Benadryl). To treat an allergic reaction.
- **Pro-pectalin.** To help with puppy diarrhea.
- **Fortiflora.** Bacterial microbes for when the puppy has diarrhea (similar to probiotics in yogurt).
- **Thermometer.** Puppies run hot; a temperature of 102°F is not uncommon, but anything higher than 103°F and they should see a vet.
- **Disinfectant wipes.** Can never have too many.
- **Gloves.**

PUPPY FUN

- **Recycled rubber balls.** Dog toys create a mountain of trash each year. We try to be responsible by buying dog toys that are made out of recycled material.
- **Chew toys.** Canine Companions puppies are only allowed to chew on the softer items, to prevent breaking their teeth. Some Nylabones, for example, are too hard (for anything our dogs chew, you must either be able to bend it a little or dent your fingernail into it).
- **Stuffies.** You cannot just dig out your kids' old stuffies and give them to your puppy. Stuffies for dogs are different than stuffies for kids. Dog stuffies do not have buttons for eyes or plastic pieces that could choke a puppy. They are generally tougher. They usually have squeakers that puppies love. We buy stuffies that are made from sturdy recycled material, like fire hoses.
- **Treats.** We chiefly use kibble as a reward, but sometimes, during training or nail clipping, we have to ramp up. The

hierarchy of treats goes: kibble, Zuke's, Little-Jacs, peanut butter, Cheez Whiz. Note: Make sure the peanut butter has no xylitol, which is toxic to dogs. Also Cheez Whiz
and peanut butter can give puppies diarrhea, so whenever
possible, we swap these with pureed pumpkin or sweet
potato—you can buy them in pouches that are usually for
babies.

- **Kong.** A Kong filled with frozen peanut butter or sweet
  potato will give you twenty minutes of free time.

Home Schedule

This is the schedule we send with anyone who has to look after a puppy at home—for example over the weekend. We understand that this schedule is not doable for people who work all day, so adjust as necessary. The most important part of the schedule is small frequent training sessions and lots of walks. All skills are in CAPS.

7:00 a.m.	**Wake up**
	• Potty break (HURRY)
	   – Usually the puppy is too excited about breakfast to poop, so just make sure they pee.
	• Breakfast (SIT, WAIT, OKAY)
	   – All meals are in the crate. Puppy should SIT, and then WAIT until you put the food down (up to thirty seconds), then OKAY means they can eat.
	• Potty break
	   –This potty break is longer, about fifteen minutes. Puppy should poop.
	
	Note about potty breaks. Until the puppy is toilet trained, they should be taken outside every thirty minutes (on the hour and half hour) during the day, and every time they wake up from a nap. They usually need to pee fifteen minutes after they drink, and poop twenty to thirty minutes after they eat.
	
	When they are eight to ten weeks old, you can carry them outside when they first wake up. It is a little known fact that puppies rarely pee or poop while they are being carried.

7:30 a.m.	**Big walk**
	• This is their big walk of the day (DRESS).

- 8 weeks = 0.5 mile
- 9 weeks = Start training on a doggy leader.
- 10 weeks = 1 mile
- 12 weeks = 1.5 miles
- 14 weeks = 2 miles
- 16 weeks = 3 miles (This is the farthest the puppy walks until twenty weeks.)

Note on walking. These walking sessions are supposed to be super fun for the puppy while they get used to walking on a doggy leader. Lots of praise and lots of treats. Every three steps if necessary. Puppy should walk at their own pace, occasionally running if they feel like it. Don't tug or jerk on the leash. If you have an older (well-trained) dog or an enthusiastic child, you can bring them to coax the puppy along. You can also walk the puppy before breakfast, so you can use their kibble as treats and they will be more motivated because they are hungry. Look for signs of fatigue like flopping down on their belly, panting, or whining, and adjust the walk accordingly. Rubbing their nose on the ground to try and get the doggy leader off does not count as fatigue.

9:00 a.m.	**Nap in crate**
	• Potty break. Try to get your puppy to pee and poop.
	• Lure your puppy into their crate with a treat (KENNEL).
	• Length of nap depends on age.
	– 8 weeks = 2 hours
	– 12 weeks = 3 hours
	– 16 weeks = 3 hours but can do 4 hours on occasion
	• Potty break
	Note on naps. The crate is a quiet place for your puppy to learn to be alone; to get a break from kids, rules, and other pets; and to relax. The crate should be in an empty, quiet room, and you should occasionally leave the house, so the puppy is *really* alone. No one should disturb the puppy while they are in their crate.
	If puppy barks for more than five minutes, and they did not pee *and* poop before their nap, take them outside for a ten-minute potty break. Then put them in their crate and ignore them for the rest of nap time.
12:00 p.m.	**Recess**
	• Lunch. Practice SIT, WAIT, OKAY before feeding.
	• Recess. Have your puppy spend some time outside if possible. Your puppy can also be inside. This is just a few hours where they can wander around, nap, or explore.
	Note on recess. Puppy must be supervised during this time. You will be amazed by how much of this time you spend pulling things out of your puppy's mouth.

2:00 p.m.	**Socialization**
	• Your goal during this time is to get the puppy to have some kind of new experience.
	    – New people (all different kinds)
	    – Dogs that you know have been fully vac-cinated, are healthy, and tolerate puppies (If you don't know the dog, no contact until twenty weeks.)
	    – New sounds, places, and things (traffic, vacuum cleaner, robotic cats)
	• This session should also double up as exercise (e.g., a twenty-minute fetch session, a short half-mile walk, a game of tug or fetch).
	
	Note on socialization. The idea is to introduce the puppy to these new experiences gradu-ally, not to flood and overwhelm them. Watch for signs of stress, like lip licking, yawning, and cringing. Calm them with praise and treats, and if they are still stressed, dial it back.
	
	You can also have people come to puppy-sit or drop your puppy off with grandparents, friends, or neighbors for a few hours a day, or a week-end. It's important that your puppy has as many positive social experiences as possible.

3:00 p.m.	**Nap in crate** • Potty break • Nap • Potty break *Note.* This nap can overlap with your dinnertime. Either way, your puppy should be in their crate while you eat, so they are not begging at the table and you don't have to watch them.
7:00 p.m.	**Dinner** • Last meal of the day.

8:00 p.m.	**Remove puppy's water** • The puppy should have easy access to fresh water during the day. Around three hours before bedtime, take away their water, so it is easier for them not to go to the bathroom in their crate. After you remove their water, no walks or frantic playtime, so they don't get too thirsty.
8:30 p.m.	**Cradle** • This is the best part of the day. Your puppy on their back while you gently rub every inch of them (CRADLE). At first you can just pick them up and lie them in the right position. But eventually they should roll into a cradle, so you can work on SIT, then DOWN, then ROLL. • Clean ears. There are medicated ear wipes to keep their ears clean—only clean the outer ears, don't go into the ear canal. • Brush teeth. Use either a dog toothbrush or a finger toothbrush and dog toothpaste. Oral hygiene is as important for dogs as it is for us. Brushing their teeth every day keeps them free from tooth and gum decay and keeps their breath smelling good. • Gently brush their hair with puppy brush. • Clip their nails carefully once a week. Shave the tiniest bit off each nail and give them a treat as soon as you clip each one. The goal is not to actually trim their nails too far, but to get puppies used to the feeling of having their nails clipped. Have some styptic powder and a cotton swab handy just in case you do accidentally cut too far. On the days you are not clipping, just touch each nail with the clippers and rub in between their toes.

10:45 p.m.	**Last potty break** • Take the puppy out for a long bathroom break—at least fifteen minutes. The goal is to get them to pee *and* poop.
11:00 p.m.	**Bedtime** • Ask them to go in their crate (KENNEL). Throw in a treat if necessary. • If the puppy cries for more than fifteen minutes, take them outside for fifteen minutes. If they have peed or pooped, put them back in their crate and ignore them. • At eight weeks old, if they sleep for more than two hours and wake up and whine, take them to the bathroom. • At twelve weeks old, your puppy should be able to spend seven to eight hours a night in their crate without going to the bathroom, unless they have diarrhea or a urinary tract infection. *Note on bedtime.* When your puppy sleeps in a crate, they can't wander off in the middle of the night and go to the bathroom somewhere for you to find in the morning.

The goal is that by twenty weeks, your puppy should be:

- crate trained
- potty trained
- able to walk with a loose leash using a doggy leader
- well-versed in eight skills (SIT, DOWN, KENNEL, NO, HURRY, DRESS, WAIT, OKAY)

Dealing with Diarrhea

Hopefully, you never need this section. But you probably will. In a survey of over 4,700 Labrador retrievers, over one in three dogs reported diarrhea at least once.[1] Most of the diarrhea occurred from eight weeks to twenty-four weeks old—overlapping almost exactly the age when the puppies are in our kindergarten. You can imagine we are fairly experienced in this area.

If an accident happens while you and the puppy are together, immediately take the puppy outside for at least fifteen minutes. Wait to clean up the mess until you are back inside (otherwise your puppy might make the mess bigger). Then, take your puppy to the bathroom every thirty minutes.

If your puppy has diarrhea in their crate after a nap or waking up in the morning, there is a clear procedure.

First, do not panic. Your impulse will be to get your

puppy out of the crate immediately. This is a mistake. Before you touch anything, walk away and assemble your supplies.

You will need:

- disinfectant
- old towels
- old rags
- paper towels
- puppy shampoo
- a change of clothes

Once your supplies are easily accessible, the second step is to secure your puppy. You need to put the puppy somewhere with the understanding that everywhere the puppy goes and everything the puppy touches from this point on will have to be cleaned and disinfected. A good idea is to create a space outside where they are safe and enclosed (this is where x-pens come in handy). For dorm volunteers and people who live in apartments, or others who have no way to secure their puppy, it makes sense to bathe the puppy immediately and then proceed to the third step.

The third step is to clean the crate. The crates we use in the Puppy Kindergarten have a tray that slides out from underneath. When the puppies came home with us, their crates were next to the bathroom for exactly this reason. You can tip the mess into the toilet and take the tray outside and spray it off with a hose.

The fourth step is to decontaminate the puppy. We usually give them a shower, just because blasting them with a hose in the middle of the night or early morning is not much fun for either of you. If you don't have one of those shower-heads attached to a hose, you might want to get one that attaches to your bathtub nozzle.

Towel off the puppy and put them in an enclosed space (again, the x-pen). At this point, you will probably need a shower. Then you need to disinfect your shower, throw the dirty towels, bath mat, pj's, stuffies, and any other collateral damage into the washing machine.

The final step is to disinfect everything. We have a gallon of industrial-grade disinfectant called Rescue for this purpose. Mop the bathroom floor; wipe door handles, the washing machine, and banisters. Splash disinfectant where you kept them outside and hose it down for good measure.

The reason we go through this extended protocol is because there are dozens of pathogens, bacteria, and parasites that might cause diarrhea. You don't want it to reinfect your puppy or any other dogs around the house.

But puppies have diarrhea for all sorts of reasons. The puppy might have eaten something that doesn't agree with them. They might have eaten too much or too fast. They might be anxious or stressed. They might have been to the vet for a vaccine, or just had their deworming medication.

A few episodes of diarrhea are nothing to worry about. However, because puppies are so young, dehydration caused by diarrhea can be dangerous. So if they have diarrhea, follow them to every bathroom break, carefully examining their poop so you can tell your vet if you see anything unusual. We look for plant matter, cloth material, blood, foreign objects, or an unusual smell or color.

Each volunteer learns the Bristol Stool Scale, aka the poop chart, by heart. The poop chart was developed by the Bristol Royal Infirmary in 1997 to measure the severity of people suffering from bowel disorders. The chart rates poop on a scale of one to seven, with one as dry pellets and seven as a liquid puddle.

BRISTOL POOP SCALE

4 Long and loose
Log-shaped but moist

1 Ouch!
Separate hard lumps
that are hard to pass

5 Still OK
Pretty moist, but
still has some shape

2 Nice form
Firm, not too hard,
looks like a cracked log

6 Uh-oh . . .
No shape, moist, and
leaves a residue

3 Perfect!
Long and log-shaped,
smooth on the surface

7 Long night ahead
Uncontrolled spurts
of liquid

Three is the magic number for diarrhea. Three poops rating a six or seven trigger diarrhea protocol, which is as follows:

- One dose of Pro-Pectalin, a nutrient that helps maintain gut health and pH, three times a day.
- One dose of probiotic, the live bacteria that are in yogurt and are good for digestive health.
- Reduce food portions by 30 percent at each meal.

Three days of diarrhea, or diarrhea plus vomiting, or diarrhea and lethargy all prompt a visit to the vet.

You Are What You Eat

At the Puppy Kindergarten, since what comes out of a puppy can be so consequential, we are equally concerned with what goes into a puppy. All Canine Companions puppies are fed high quality kibble throughout their service. Puppy Kindergarten volunteers can eat in front of the puppies—it is good training for the puppies to be around human food—but they are not allowed to feed the puppies human food. Service dogs must be able to lie quietly under a restaurant table without drooling on the diners. Kibble is convenient because it does not make a mess like wet food. It is easy to measure out, so if the puppies eat a lot of kibble during class, we can measure out the remainder for their lunch. And when there are a hundred and fifty people taking care of seven puppies, it is important to keep track.

For dogs who are not service dogs, we feed them as dogs were fed for thousands of years, with scraps from our table. Our pet dogs eat curries, salads, and fruit. They eat the children's leftover crusts, food we drop on the floor while preparing dinner, and they lick the plates afterward. A dog's diet should be supplemented with kibble, but with a careful eye on their weight and plenty of exercise, dogs can be an excellent control on the food waste of our household. This lowers our carbon paw print and is better for the environment.

Margaret says table scraps are perfect for dogs because presumably anything that you have eaten is free from harmful pathogens or bacteria. There are a few foods that are poisonous to dogs: grapes and grape products, onions, garlic, beer, alcohol, and xylitol are the main ones. Xylitol is tricky since it is in different foods like artificial sweeteners and some peanut butters. Check with your vet for a comprehensive list.

We also limit table scraps to dinner scraps. No desserts or anything with chocolate or sugar. No bones, of course, as these can splinter. But in general, anything we eat, our dogs can eat too. We avoid the raw food diet based on a number of research findings that concluded it has a bunch of negative consequences for dogs, including a higher rate of E. coli and salmonella, as well as more serious effects like nutritional osteodystrophy—a bone disease in puppies—as well as thyroid disorders.[2]

NOTES

Chapter 1: The Puppy Brain

Page 7   **By their six-week mark, puppies can see**   Dogs have what is called "scotopic vision," meaning they are able to see in low light.[1] In studies of free-ranging dogs (dogs who live without owners), their most active time for foraging and hunting is at dawn and dusk.

It is often said dogs are color blind and can only see in shades of gray. In reality dogs can see colors; they just make out fewer colors than we do as primates. Dogs can see blue and yellow—it is red and green they have trouble seeing, as do color-blind people.[2]

Dogs also have lower visual acuity than humans, which means the images they see are not as clear. For example, an object you can see at seventy-five feet away, a dog can only see once it is twenty feet away.[3] Dogs also have a higher flicker fusions rate, which is the rate at which a flickering image tends to merge into a smooth image. So, when a dog is watching a standard television or computer screen, while you see smooth movement, dogs see a flickering, stuttering image that probably does not make much sense.[4]

Even when dogs do see video specially formatted for them, it is not clear how well they understand them. For example, if your dog sees you on a video call, even though the image is played at the correct speed, there is not much evidence they understand that the image actually represents you. When dogs see a photo or a video of something they usually associate with a reward, like a toy, they show limited activity in the reward-processing regions of their brains. This may indicate that most dogs do not automatically understand that the two-dimensional image is a representation of the three-dimensional, real-life version.[5]

Page 9   **why it is believed to be the critical period of socialization**   This critical period is in part inferred from work with other mammals. In rodents there is a critical period for myelination of the medial prefrontal cortex (mPFC). This region of the brain is responsible for the most complex forms of social behavior involving inhibition and memory. Social experiences are necessary during the juvenile period, just after weaning, for normal development to occur in this region. Mice deprived of appropriate social stimulation during the post-weaning period do not show normal myelination of mPFC even after being placed in an enriched social environment as adults.[6]

Chapter 3: Celebrating Individuality

Page 32 **which of the dogs were most likely to successfully complete their training** Dogs offer us a great opportunity to distinguish between general intelligence theory and multiple intelligence theory. Together with anthropologist Evan MacLean, we created the largest set of cognitive games that had ever been developed for dogs. We included every cognitive test we could think of and ended up with a battery of twenty-five games. It took each dog an hour per day for four days to complete all the games. We tested over a hundred adult pet dogs to make sure what we learned could be applied to everyone's pet dog. We also tested over two hundred service dogs and two hundred bomb-detecting dogs just before they were professionally trained at around eighteen months of age. This allowed us to look at the relationship between performance on all the different problems we presented and their subsequent success or failure with professional training. If dogs simply have some type of general intelligence, like "learning ability" or "IQ," then performance in all the different games should be equally related to one another. We could then easily rank dogs as more smart or less smart. But that is not what we found. Instead, performance in games that tested similar cognitive skills tended to cluster together. If a dog performed well on one type of memory game they probably performed well on other memory games, and if they performed well on one communication game they did well on the other communication games, and so on. However, good memory did not mean you were a high-performing communicator and vice versa. Overall, our analysis revealed at least five, and probably six, different types of cognitive abilities in dogs. Cognition is a major player in generating individuality in our dogs. Each of the cognition types is like a letter in an alphabet that can be combined and recombined in different ways. This recombination shapes personalities and gives different, perhaps unique, skills, biases, and weaknesses to individual dogs. Instead of dog intelligence being a one-dimensional sliding scale where a dog can be ranked from smart to not so smart, dogs have a wide variety of cognitive profiles. With further analysis we found that performance on some of our cognitive games was linked with training success in both the service and military dogs. Of the twenty-five cognitive games, there were eleven that were most strongly associated with professional training success, including the ability to make eye contact, make inferences, follow a human pointing gesture, and exercise self-control. Our next step was to test another large group of service dogs with just these eleven games. Based on their results, we made predictions on who would make it through training and who would not. Our predictions were up to 90 percent accurate. Our cognitive data was particularly good at predicting which dogs were *most* likely to make it. We finally had a set of games that an adult dog could play in under an hour that would, with high accuracy, predict their success.[7]

- **Senses.** The first cognitive domain is the senses, so we test the puppies to make sure they can see, hear, and smell.

- **Temperament.** The second domain is temperament, where we test how they react to strange people and even stranger toys.

- **Social.** The third domain is social skills; we test the basic theory of mind abilities related to cooperative communication. This includes games with several levels of difficulty to test how well the puppies understand gestures. We begin with the marker test we introduced with Ashton, Aiden, and Dune, which most puppies seem to pass easily, then progress to a brief pointing gesture, which most puppies find difficult. We also include a game of fetch, because bringing back a toy for joint play indicates a willingness to cooperate and the ability to ask for help.

- **Executive.** The fourth cognitive domain refers to abilities known as executive function. These are cognitive skills required to solve almost every type of problem. Our tests include measures of memory and self-control.

- **Physical.** The final cognitive domain is what we call "physics" because it refers to how puppies understand the properties of the physical world. Our games measure if puppies understand solidity enough to infer where a treat is hidden. This requires the knowledge that solid objects cannot pass through one another.

Chapter 6: Brainy Genes

Page 75 **avenue to increase the supply of service dogs** We first attempted to find breed differences in cognition in 2009.[8] Tory Wobber wanted to compare working dogs to non-working dogs in their ability to read the cooperative-communicative gestures of humans. Around this same time, genetic sequencing had revealed that while most dog breeds appeared in the past few hundred years, there are a few breeds that are genetically more wolflike, or ancient. Since wolves do not follow the cooperative-communicative gestures of humans like dogs do, Tory compared wolflike breeds to modern breeds on this ability.

In the end, she settled on four breeds.

	Wolflike breeds	**Modern breeds**
Working dogs	**Siberian huskies** One of the eleven genetically wolflike breeds, they have been intensively bred as working dogs, transporting goods and people over frozen landscapes, first in Siberia, then later Alaska.	**German shepherds** A modern breed descended from sheep-herding dogs in Germany, they became popular in the United States after the show *Rin Tin Tin*, but have since been heavily selected as working dogs in a range of different jobs, including military, police, and search and rescue.

	Wolflike breeds	Modern breeds
Non-working dogs	**Basenjis** Another wolflike breed, they are probably representative of what the first domesticated dogs looked like. Basenjis are depicted in art from ancient Babylonia and Mesopotamia, and were presented as gifts to Egyptian pharaohs. When these ancient civilizations fell, basenjis remained, semi wild, along the banks of the Nile and Congo rivers. Congolese huntsmen especially prize basenjis as hunting dogs, but basenjis are not considered working dogs because they chase down and quarry their prey independently, without any help from their owners.	**Toy poodles** They are the quintessential modern dogs. While standard poodles were originally working water dogs, toy poodles were bred for no work at all in the 1900s, to fit into the city lifestyles and apartments of the new middle class.

Tory used the same games to test each of these four breeds for their ability to read the cooperative-communicative cues of humans. When she compared the results, she found that if dogs had been bred to work with humans, they were better at following cooperative-communicative gestures than dogs who were not bred to work with humans, regardless of how wolflike the breed was. This indicated that this one cognitive skill could be selected for in a breed, and therefore might have a genetic component.

Other researchers found that border collies were better than Anatolian shepherds at following cooperative-communicative gestures, but the Anatolian shepherds improved with experience.[9]

Chapter 8: Critical Experience

Page 116 **what happens to animals who grow up without it** Human babies who are neglected exhibit abnormal behavior even into adulthood. They have trouble regulating their emotions and forming attachments. Abused children show an altered pattern of resting brain activity and a reduced hippocampus.[10] Orphans who suffered severe neglect in institutions have smaller than average brains with less developed white and gray matter in their cerebral cortex. These orphans have different baseline brain activity than other children and difficulty recognizing and processing emotions on people's faces. People need social experiences, not just from caregivers but from many different people. The more people you have positive interactions with as a baby, the more tolerant you are as an adult. The more tolerant you are, the better at solving problems you become, because you can partner, cooperate, and learn from others.

The same is true for puppies and other animals.

Page 123 **impact of early rearing experiences** All
puppy experiences begin with their mothers. Born blind,
deaf, and with a poor sense of smell, puppies are completely
dependent on her for weeks. They need her to regulate their
temperature and cannot even pee or poop until she licks
their genitals. Their mother is often a single parent. Dogs
are not cooperative breeders like wolves. A wolf female with
puppies can count on their father to take care of them while
she rests or hunts. But domestication has altered the psychology of male dogs, and fathers provide little help to mothers in rearing the puppies.

Free-ranging dogs, also called village dogs or feral dogs,
are any group of dogs who do not rely on human interven-
tion to survive. Free-ranging dogs exist everywhere people
are found. They roam around neighborhoods, scavenging
in trash cans and garbage dumps. They do not go home at
night and do not have owners. They sleep in ditches, rock
cavities, or storm drains. Although unrelated males might
play with puppies and sometimes regurgitate food for them,
dog fathers still provide little, if any, care or provisioning for
their puppies. If dog dads contribute, they mostly just guard
the den.[11]

Page 125 **lower levels of aggression, anxiety, and
fear** Multiple studies indicate that attentive maternal care
offers beneficial effects. Puppies isolated for even a short
period of time were at risk for developing abnormal fear responses, social inadequacies, hyperactivity, increased aggression, decreased learning ability, separation anxiety, and

increased stress behaviors.[12] The Swedish Armed Forces found that puppies with attentive mothers were less fearful, were more likely to engage with their social and physical world, and had lower levels of aggression, anxiety, and fear.[13] In another study, eight-week-old beagle puppies who had less attentive mothers were more likely to show destructive behaviors to their enclosure and call out in distress during an isolation test. In contrast, puppies with greater maternal care showed increased engagement with the environment and reduced signs of distress.[14]

Page 126   **daily handling exercises, or "gentling"**   This study was a replication of a secret program conducted over half a century ago. During the Vietnam War, the U.S. government started a breeding and training program to deploy dogs that could support U.S. troops in the field. The project was known as "Super Dog," and the objective was to create a kind of hero dog, resistant to stress with superior problem-solving abilities. The methods consisted of manipulating a puppy's earliest experiences.[15] There are no public records from Project Super Dog, but the dogs deployed to Vietnam were so successful that enemy soldiers ranked the dogs as more dangerous than U.S. troops, even though the troops carried weapons and the dogs did not.

Chapter 10: Memory at Work

Page 145   **whether a dog remembers us if we leave for a long time**   There is also evidence that dogs can remember a vast amount of information. Researchers demonstrated that dogs can perform fast mapping, the same cognitive process that children use to remember new words and learn languages.[16] Chaser could remember the names for over a thousand objects. Once Chaser learned the names of the objects, she remembered all of them, even a month later. In total, Chaser remembered the names of 1,022 toys, stuffies, Frisbees, and balls. The way she learned the names was remarkable. For example, say you scattered ten toys around the room, which included nine toys she knew the names of and one toy Smurf she had never seen before. Then you asked Chaser to "go get the Smurf." Chaser would go straight to the Smurf, even though she had never seen the toy before, nor heard the word "Smurf." This is because Chaser remembered the names of the nine toys that were familiar to her and realized that "Smurf"—a word she had never heard—referred to the toy she had never seen. This cognitive process is called "fast mapping," and it is the same process that children use to remember new words and learn language.

When researchers played dogs a recording of the voice of their owner or that of a stranger, then followed up by showing them a photo, dogs had formed an expectation about the picture they would see based on the voice they heard.[17]

Chapter 11: Our Takeaways

Page 161 **an overall pattern emerges** Out of a test battery of eleven games, nine showed a tendency for puppy and adult success to be related.[18] But five of these games showed a *really* strong signal of stability from puppy to adult performance:

- interest in looking to humans
- persistence in solving an unsolvable task
- self-control with the cylinder
- odor discrimination
- communicative marker (and almost pointing)

Now we have a set of games that we can play with puppies that helps predict what they will do as adults. These games are specifically designed to measure the skills that are related to their training success as adults, but the advantage of these games is that they do not require any time-intensive training to complete.

REFERENCES

Introduction: Congo

1. B. Hare, "From Hominoid to Hominid Mind: What Changed and Why?" *Annual Review of Anthropology* 40 (2011), 293–309.
2. B. Hare, "Survival of the Friendliest: Homo Sapiens Evolved via Selection for Prosociality." *Annual Review of Psychology* 68 (2017), 155–186.
3. B. Hare, V. Woods, *The Genius of Dogs: How Dogs Are Smarter Than You Think.* (Penguin, 2013).
4. E. E. Bray, M. E. Gruen, G. E. Gnanadesikan, D. J. Horschler, K. M. Levy, B. S. Kennedy, B. A. Hare, E. L. MacLean, "Dog Cognitive Development: A Longitudinal Study across the First 2 Years of Life." *Animal Cognition* 24 (2021), 311–328.
5. E. E. Bray, K. M. Levy, B. S. Kennedy, D. L. Duffy, J. A. Serpell, E. L. MacLean, "Predictive Models of Assistance Dog Training Outcomes Using the Canine Behavioral Assessment and Research Questionnaire and a Standardized Temperament Evaluation." *Frontiers in Veterinary Science* 6 (2019), 49.
6. B. Hare, M. Ferrans, "Is Cognition the Secret to Working Dog Success?" *Animal Cognition* 24 (2021), 231–237.

7. E. L. MacLean, B. Hare, "Enhanced Selection of Assistance and Explosive Detection Dogs Using Cognitive Measures." *Frontiers in Veterinary Science* (2018), 236.

Chapter 1: The Puppy Brain

1. B. Hare, V. Woods, *Survival of the Friendliest: Understanding Our Origins and Rediscovering Our Common Humanity*. (Random House Trade Paperbacks, 2021).
2. D. Jardim-Messeder, K. Lambert, S. Noctor, F. M. Pestana, M. E. de Castro Leal, M. F. Bertelsen, A. N. Alagaili, O. B. Mohammad, P. R. Manger, S. Herculano-Houzel, "Dogs Have the Most Neurons, Though Not the Largest Brain: Trade-off between Body Mass and Number of Neurons in the Cerebral Cortex of Large Carnivoran Species." *Frontiers in Neuroanatomy* 11 (2017), 118.
3. R. J. Ruben, "The Ontogeny of Human Hearing." *Acta Oto-Laryngologica* 112 (1992), 192–196.
4. K. Lord, "A Comparison of the Sensory Development of Wolves (Canis lupus lupus) and Dogs (Canis lupus familiaris)." *Ethology* 119 (2013), 110–120.
5. M. Fox, "Neuronal Development and Ontogeny of Evoked Potentials in Auditory and Visual Cortex of the Dog." *Electroencephalography and Clinical Neurophysiology* 24 (1968), 213–226.
6. T. Jezierski, J. Ensminger, L. Papet, *Canine Olfaction Science and Law: Advances in Forensic Science, Medicine, Conservation, and Environmental Remediation* (CRC Press, 2016).
7. M. W. Fox, "Postnatal Growth of the Canine Brain and Correlated Anatomical and Behavioral Changes during Neuro Ontogenesis." *Growth* 28 (1964), 135–141.
8. M. Fox, "Gross Structure and Development of the Canine Brain." *American Journal of Veterinary Research* 24 (1963), 1240–1247.
9. B. Gross, D. Garcia-Tapia, E. Riedesel, N. M. Ellinwood, J. K. Jens, "Normal Canine Brain Maturation at Magnetic Resonance Imaging." *Veterinary Radiology & Ultrasound* 51 (2010), 361–373.

Chapter 2: Ready for School

1. B. Hare, "From Hominoid to Hominid Mind: What Changed and Why?" *Annual Review of Anthropology* 40 (2011), 293–309.
2. M. Tomasello, *Becoming Human: A Theory of Ontogeny*. (Harvard University Press, 2019).
3. Hare, "From Hominoid to Hominid Mind." 293–309.
4. B. Hare, M. Ferrans, "Is Cognition the Secret to Working Dog Success?" *Animal Cognition* 24 (2021), 231–237.
5. B. Hare, V. Woods, *The Genius of Dogs: How Dogs Are Smarter than You Think*. (Penguin, 2013).
6. Hare, Ferrans, "Is Cognition the Secret to Working Dog Success?" 231–237.
7. B. Hare, J. Call, M. Tomasello, "Communication of Food Location between Human and Dog (Canis familiaris)." *Evolution of Communication* 2 (1998), 137–159.
8. F. Rossano, M. Nitzschner, M. Tomasello, "Domestic Dogs and Puppies Can Use Human Voice Direction Referentially." *Proceedings of the Royal Society B: Biological Sciences* 281 (2014), 20133201.
9. B. Hare, M. Tomasello, "Domestic Dogs (Canis familiaris) Use Human and Conspecific Social Cues to Locate Hidden Food." *Journal of Comparative Psychology* 113 (1999), 173.
10. E. L. MacLean, E. Herrmann, S. Suchindran, B. Hare, "Individual Differences in Cooperative Communicative Skills Are More Similar Between Dogs and Humans than Chimpanzees." *Animal Behaviour* 126 (2017), 41–51.
11. Hare, Ferrans, "Is Cognition the Secret to Working Dog Success?" 231–237.
12. Hare, Tomasello, "Domestic Dogs (Canis familiaris) Use Human and Conspecific Social Cues." 173.
13. B. Hare, M. Brown, C. Williamson, M. Tomasello, "The Domestication of Social Cognition in Dogs." *Science* 298 (2002), 1634–1636.
14. H. Salomons, K. C. Smith, M. Callahan-Beckel, M. Callahan, K. Levy, B. S. Kennedy, E. E. Bray, G. E. Gnanadesikan, D. J. Horschler, M. Gruen, "Cooperative Communication with Humans Evolved to Emerge Early in Domestic Dogs." *Current Biology* 31(2021), 3137–3144, e3111.

15. J. Riedel, K. Schumann, J. Kaminski, J. Call, M. Tomasello, "The Early Ontogeny of Human–Dog Communication." *Animal Behaviour* 75 (2008), 1003–1014.
16. MacLean, Herrmann, Suchindran, Hare, "Individual Differences in Cooperative Communicative Skills." 41–51.

Chapter 3: Celebrating Individuality

1. E. E. Bray, K. M. Levy, B. S. Kennedy, D. L. Duffy, J. A. Serpell, E. L. MacLean, "Predictive Models of Assistance Dog Training Outcomes Using the Canine Behavioral Assessment and Research Questionnaire and a Standardized Temperament Evaluation." *Frontiers in Veterinary Science* 6 (2019), 49.
2. E. M. Macphail, J. J. Bolhuis, "The Evolution of Intelligence: Adaptive Specializations versus General Process." *Biological Reviews* 76 (2001), 341–364.
3. E. L. MacLean, E. Herrmann, S. Suchindran, B. Hare, "Individual Differences in Cooperative Communicative Skills Are More Similar between Dogs and Humans than Chimpanzees." *Animal Behaviour* 126 (2017), 41–51.
4. MacLean, Herrmann, Suchindran, Hare, "Individual Differences in Cooperative Communicative Skills." 41–51.
5. L. Lazarowski, S. Krichbaum, L. P. Waggoner, J. S. Katz, "The Development of Problem-solving Abilities in a Population of Candidate Detection Dogs (Canis familiaris)." *Animal Cognition* 23 (2020), 755–768.
6. A. Horowitz, "Disambiguating the 'Guilty Look': Salient Prompts to a Familiar Dog Behaviour." *Behavioural Processes* 81 (2009), 447–452.
7. J. Kaminski, B. M. Waller, R. Diogo, A. Hartstone-Rose, A. M. Burrows, "Evolution of Facial Muscle Anatomy in Dogs." *Proceedings of the National Academy of Sciences* 116 (2019), 14677–14681.
8. B. M. Waller, K. Peirce, C. C. Caeiro, L. Scheider, A. M. Burrows, S. McCune, J. Kaminski, "Paedomorphic Facial Expressions Give Dogs a Selective Advantage." *PLoS One* 8 (2013), e82686.
9. E. E. Bray, M. E. Gruen, G. E. Gnanadesikan, D. J. Horschler, K. M. Levy, B. S. Kennedy, B. A. Hare, E. L. MacLean, "Dog

Cognitive Development: A Longitudinal Study across the First 2 Years of Life." *Animal Cognition* 24 (2021), 311–328.

Chapter 4: Control Yourself

1. Y. Shoda, W. Mischel, P. K. Peake, "Predicting Adolescent Cognitive and Self-regulatory Competencies from Preschool Delay of Gratification: Identifying Diagnostic Conditions." *Developmental Psychology* 26 (1990), 978.
2. Falk, A., Kosse, F., Pinger, P. (2020). *Re-Revisiting the Marshmallow Test: A Direct Comparison of Studies* by Shoda, Mischel, and Peake (1990), and Watts, Duncan, and Quan (2018). *Psychological Science,* 31(1), 100–104.
3. J. P. Tangney, R. F. Baumeister, A. L. Boone, "High Self-control Predicts Good Adjustment, Less Pathology, Better Grades, and Interpersonal Success." *Journal of Personality* 72 (2004), 271–324.
4. E. L. MacLean, B. Hare, C. L. Nunn, E. Addessi, F. Amici, R. C. Anderson, F. Aureli, J. M. Baker, A. E. Bania, A. M. Barnard, "The Evolution of Self-control." *Proceedings of the National Academy of Sciences* 111 (2014), E2140–E2148.
5. E. L. MacLean, B. Hare, "Enhanced Selection of Assistance and Explosive Detection Dogs Using Cognitive Measures." *Frontiers in Veterinary Science* (2018), 236.
6. N. J. Rooney, S. A. Gaines, J. W. S. Bradshaw, S. Penman, "Validation of a Method for Assessing the Ability of Trainee Specialist Search Dogs." *Applied Animal Behaviour Science* 103 (2007), 90–104.

Chapter 5: A Surprising Marker of Success

1. B. L. Hart, L. A. Hart, A. P. Thigpen, A. Tran, M. J. Bain, "The Paradox of Canine Conspecific Coprophagy." *Veterinary Medicine and Science* 4 (2018), 106–114.
2. B. Boze, "A Comparison of Common Treatments for Coprophagy in Canis familiaris." *Journal of Applied Companion Animal Behavior* 2 (2008), 22–28.
3. E. E. Bray, K. M. Levy, B. S. Kennedy, D. L. Duffy, J. A. Serpell,

E. L. MacLean, "Predictive Models of Assistance Dog Train-
ing Outcomes Using the Canine Behavioral Assessment and
Research Questionnaire and a Standardized Temperament
Evaluation." *Frontiers in Veterinary Science* 6 (2019), 49.

4. E. Raffan, R. J. Dennis, C. J. O'Donovan, J. M. Becker, R. A.
    Scott, S. P. Smith, D. J. Withers, C. J. Wood, E. Conci, D. N.
    Clements, "A Deletion in the Canine POMC Gene Is As-
    sociated with Weight and Appetite in Obesity-prone Labrador
    Retriever Dogs." *Cell Metabolism* 23 (2016), 893–900.

5. B. Osthaus, S. E. Lea, A. M. Slater, "Dogs (Canis lupus familia-
    ris) Fail to Show Understanding of Means-end Connections in a
    String-Pulling Task." *Animal Cognition* 8 (2005), 37–47.

6. Osthaus, Lea, Slater, "Dogs (Canis lupus familiaris) Fail to
    Show Understanding." 37–47.

7. E. L. MacLean, B. Hare, "Enhanced Selection of Assistance and
    Explosive Detection Dogs Using Cognitive Measures." *Frontiers
    in Veterinary Science* (2018), 236.

Chapter 6: Brainy Genes

1. E. L. MacLean, B. Hare, "Enhanced Selection of Assistance and
    Explosive Detection Dogs Using Cognitive Measures." *Frontiers
    in Veterinary Science* (2018), 236.

2. E. A. Ostrander, A. Ruvinsky, *The Genetics of the Dog.* (CABI,
    2012).

3. A. R. Boyko, P. Quignon, L. Li, J. J. Schoenebeck, J. D. Degen-
    hardt, K. E. Lohmueller, K. Zhao, A. Brisbin, H. G. Parker,
    B. M. Vonholdt, "A Simple Genetic Architecture Underlies Mor-
    phological Variation in Dogs." *PLoS Biology* 8 (2010), e1000451.

4. E. Turkheimer, "Three Laws of Behavior Genetics and What
    They Mean." *Current Directions in Psychological Science*
    9 (2000), 160–164.

5. N. G. Gregory, T. Grandin, *Animal Welfare and Meat Science.*
    (CABI, 1998).

6. H. Ritvo, "Pride and Pedigree: The Evolution of the Victorian
    Dog Fancy." *Victorian Studies* 29 (1986), 227–253.

7. Ritvo, "Pride and Pedigree," 227–253.

8. Ritvo, "Pride and Pedigree," 227–253.

9. J. Gayon, "From Mendel to Epigenetics: History of Genetics." *Comptes Rendus Biologies* 339 (2016), 225–230.

10. N. B. Sutter, C. D. Bustamante, K. Chase, M. M. Gray, K. Zhao, L. Zhu, B. Padhukasahasram, E. Karlins, S. Davis, P. G. Jones, "A Single IGF1 Allele Is a Major Determinant of Small Size in Dogs." *Science* 316 (2007), 112–115.

11. H. G. Parker, K. Chase, E. Cadieu, K. G. Lark, E. A. Ostrander, "An Insertion in the RSPO2 Gene Correlates with Improper Coat in the Portuguese Water Dog." *Journal of Heredity* 101 (2010), 612–617.

12. S. Coren, *The Intelligence of Dogs: A Guide to the Thoughts, Emotions, and Inner Lives of Our Canine Companions* (Simon and Schuster, 2006).

13. V. Wobber, B. Hare, J. Koler-Matznick, R. Wrangham, M. Tomasello, "Breed Differences in Domestic Dogs' (Canis familiaris) Comprehension of Human Communicative Signals." *Interaction Studies* 10 (2009), 206–224.

14. L. Stewart, E. L. MacLean, D. Ivy, V. Woods, E. Cohen, K. Rodriguez, M. McIntyre, S. Mukherjee, J. Call, J. Kaminski, "Citizen Science as a New Tool in Dog Cognition Research." *PloS One* 10 (2015), e0135176.

15. Stewart, MacLean, Ivy, Woods, Cohen, Rodriguez, McIntyre, Mukherjee, Call, Kaminski, "Citizen Science as a New Tool." e0135176.

16. Stewart, MacLean, Ivy, Woods, Cohen, Rodriguez, McIntyre, Mukherjee, Call, Kaminski, "Citizen Science as a New Tool." e0135176.

17. D. J. Horschler, B. Hare, J. Call, J. Kaminski, Á. Miklósi, E. L. MacLean, "Absolute Brain Size Predicts Dog Breed Differences in Executive Function." *Animal Cognition* 22 (2019), 187–198; published online Epub2019/03/01 (10.1007/s10071-018-01234-1).

18. Horschler, Hare, Call, Kaminski, Miklósi, MacLean, "Absolute Brain Size Predicts." 187–198.

19. G. E. Gnanadesikan, B. Hare, N. Snyder-Mackler, E. L. MacLean, "Estimating the Heritability of Cognitive Traits across Dog Breeds Reveals Highly Heritable Inhibitory Control and Communication Factors." *Animal Cognition* 23 (2020), 953–964.

20. H. G. Parker, D. L. Dreger, M. Rimbault, B. W. Davis, A. B. Mullen, G. Carpintero-Ramirez, E. A. Ostrander, "Genomic Analyses Reveal the Influence of Geographic Origin, Migration, and Hybridization on Modern Dog Breed Development." *Cell Reports* 19 (2017), 697–708.

21. Gnanadesikan, Hare, Snyder-Mackler, MacLean, "Estimating the Heritability of Cognitive Traits." 953–964.

22. G. E. Gnanadesikan, B. Hare, N. Snyder-Mackler, J. Call, J. Kaminski, Á. Miklósi, E. L. MacLean, "Breed Differences in Dog Cognition Associated with Brain-Expressed Genes and Neurological Functions." *Integrative and Comparative Biology* 60 (2020), 976–990.

Chapter 7: A Big Personality

1. D. Rettew, *Child Temperament: New Thinking about the Boundary between Traits and Illness* (W. W. Norton & Company, 2013).

2. J. Kagan, N. Snidman, *The Long Shadow of Temperament.* (Harvard University Press, 2009).

3. Rettew, *Child Temperament.*

4. Rettew, *Child Temperament.*

5. E. L. MacLean, L. R. Gesquiere, M. E. Gruen, B. L. Sherman, W. L. Martin, C. S. Carter, "Endogenous Oxytocin, Vasopressin, and Aggression in Domestic Dogs." *Frontiers in Psychology* 8 (2017), 1613.

6. MacLean, Gesquiere, Gruen, Sherman, Martin, Carter, "Endogenous Oxytocin." 1613.

7. MacLean, Gesquiere, Gruen, Sherman, Martin, Carter, "Endogenous Oxytocin." 1613.

8. E. L. MacLean, B. Hare, "Dogs Hijack the Human Bonding Pathway." *Science* 348 (2015), 280–281.

9. N. Marsh, A. A. Marsh, M. R. Lee, R. Hurlemann, "Oxytocin and the Neurobiology of Prosocial Behavior." *The Neuroscientist* 27 (2021), 604–619.

10. MacLean, Hare, "Dogs Hijack the Human Bonding Pathway," 280–281.

11. Marsh, Marsh, Lee, Hurlemann, "Oxytocin and the Neurobiology." 604–619.

12. M. Nagasawa, T. Kikusui, T. Onaka, M. Ohta, "Dog's Gaze at Its Owner Increases Owner's Urinary Oxytocin During Social Interaction." *Hormones and Behavior* 55 (2009), 434–441.
13. S. C. Miller, C. C. Kennedy, D. C. DeVoe, M. Hickey, T. Nelson, L. Kogan, "An Examination of Changes in Oxytocin Levels in Men and Women Before and After Interaction with a Bonded Dog." *Anthrozoös* 22 (2009), 31–42.
14. B. M. Waller, K. Peirce, C. C. Caeiro, L. Scheider, A. M. Burrows, S. McCune, J. Kaminski, "Paedomorphic Facial Expressions Give Dogs a Selective Advantage." *PLoS One* 8 (2013), e82686.
15. E. L. MacLean, L. R. Gesquiere, N. R. Gee, K. Levy, W. L. Martin, C. S. Carter, "Effects of Affiliative Human–Animal Interaction on Dog Salivary and Plasma Oxytocin and Vasopressin." *Frontiers in Psychology* 1606 (2017).
16. Waller, Peirce, Caeiro, Scheider, Burrows, McCune, Kaminski, "Paedomorphic Facial Expressions Give Dogs a Selective Advantage." e82686.
17. Nagasawa, Kikusui, Onaka, Ohta, "Dog's Gaze at Its Owner Increases Owner's Urinary Oxytocin during Social Interaction." 434–441.
18. F. W. Nicholas, E. R. Arnott, P. D. McGreevy, "Hybrid Vigour in Dogs?" *The Veterinary Journal* 214 (2016), 77–83.
19. J. Yordy, C. Kraus, J. J. Hayward, M. E. White, L. M. Shannon, K. E. Creevy, D. E. Promislow, A. R. Boyko, "Body Size, Inbreeding, and Lifespan in Domestic Dogs." *Conservation Genetics* 21 (2020), 137–148.

Chapter 8: Critical Experience

1. K. Lord, "A Comparison of the Sensory Development of Wolves (Canis lupus lupus) and Dogs (Canis lupus familiaris)." *Ethology* 119 (2013), 110–120.
2. L. N. Trut, "Early Canid Domestication: The Farm-Fox Experiment: Foxes Bred for Tamability in a 40-year Experiment Exhibit Remarkable Transformations that Suggest an Interplay between Behavioral Genetics and Development." *American Scientist* 87 (1999), 160–169.
3. E. E. Bray, M. D. Sammel, D. L. Cheney, J. A. Serpell, R. M.

Seyfarth, "Characterizing Early Maternal Style in a Population of Guide Dogs." *Frontiers in Psychology* 8 (2017), 175.

4. E. E. Bray, M. D. Sammel, D. L. Cheney, J. A. Serpell, R. M. Seyfarth, "Effects of Maternal Investment, Temperament, and Cognition on Guide Dog Success." *Proceedings of the National Academy of Sciences* 114 (2017), 9128–9133.

5. P. Foyer, E. Wilsson, P. Jensen, "Levels of Maternal Care in Dogs Affect Adult Offspring Temperament." *Scientific Reports* 6 (2016), 19253.

6. H. Vaterlaws-Whiteside, A. Hartmann, "Improving Puppy Behavior Using a New Standardized Socialization Program." *Applied Animal Behaviour Science* 197 (2017), 55–61.

7. A. Gazzano, C. Mariti, L. Notari, C. Sighieri, E. A. McBride, "Effects of Early Gentling and Early Environment on Emotional Development of Puppies." *Applied Animal Behaviour Science* 110 (2008), 294–304.

Chapter 9: Bringing the Puppies Home

1. N. Bunford, V. Reicher, A. Kis, Á. Pogány, F. Gombos, R. Bódizs, M. Gácsi, "Differences in Pre-sleep Activity and Sleep Location Are Associated with Variability in Daytime/Nighttime Sleep Electrophysiology in the Domestic Dog." *Scientific Reports* 8 (2018), 1–10.

2. R. Bódizs, A. Kis, M. Gácsi, J. Topál, "Sleep in the Dog: Comparative, Behavioral and Translational Relevance." *Current Opinion in Behavioral Sciences* 33 (2020), 25–33.

3. G. J. Adams, K. Johnson, "Sleep-Wake Cycles and Other Nighttime Behaviours of the Domestic Dog Canis familiaris." *Applied Animal Behaviour Science* 36 (1993), 233–248.

4. Bunford, Reicher, Kis, Pogány, Gombos, Bódizs, Gácsi, "Differences in Pre-sleep Activity and Sleep Location Are Associated with Variability in Daytime/Nighttime Sleep Electrophysiology in the Domestic Dog." 1–10.

5. D. Oudiette, M.-J. Dealberto, G. Uguccioni, J.-L. Golmard, M. Merino-Andreu, M. Tafti, L. Garma, S. Schwartz, I. Arnulf, "Dreaming without REM Sleep." *Consciousness and Cognition* 21 (2012), 1129–1140.

6. M. Fox, G. Stanton, "A Developmental Study of Sleep and Wakefulness in the Dog." *Journal of Small Animal Practice* 8 (1967), 605–611.
7. P. M. Miller, M. L. Commons, "The Benefits of Attachment Parenting for Infants and Children: A Behavioral Developmental View." *Behavioral Development Bulletin* 16 (2010), 1.
8. Miller, Commons, "The Benefits of Attachment Parenting." 1.

Chapter 10: Memory at Work

1. L. R. Squire, B. Knowlton, G. Musen, "The Structure and Organization of Memory." *Annual Review of Psychology* 44 (1993), 453–495.
2. E. T. Rolls, "Memory Systems in the Brain." *Annual Review of Psychology* 51(2000), 599–630.
3. N. Cowan, "What Are the Differences between Long-term, Short-term, and Working Memory?" *Progress in Brain Research* 169 (2008), 323–338.
4. Squire, Knowlton, Musen, "The Structure and Organization of Memory." 453–495.
5. P. G. Hepper, "Long-term Retention of Kinship Recognition Established during Infancy in the Domestic Dog." *Behavioural Processes* 33 (1994), 3–14.
6. E. L. MacLean, B. Hare, "Enhanced Selection of Assistance and Explosive Detection Dogs Using Cognitive Measures." *Frontiers in Veterinary Science* (2018), 236.
7. G. Martin-Ordas, J. Call, "Memory Processing in Great Apes: The Effect of Time and Sleep." *Biology Letters* 7 (2011), 829–832.
8. I. B. Iotchev, A. Kis, R. Bódizs, G. van Luijtelaar, E. Kubinyi, "EEG Transients in the Sigma Range during Non-REM Sleep Predict Learning in Dogs." *Scientific Reports* 7 (2017), 1–11.
9. R. Bódizs, A. Kis, M. Gácsi, J. Topál, "Sleep in the Dog: Comparative, Behavioral and Translational Relevance." *Current Opinion in Behavioral Sciences* 33 (2020), 25–33.
10. K. Louie, M. A. Wilson, "Temporally Structured Replay of Awake Hippocampal Ensemble Activity during Rapid Eye Movement Sleep." *Neuron* 29 (2001), 145–156.
11. D. J. Foster, M. A. Wilson, "Reverse Replay of Behavioural Se-

quences in Hippocampal Place Cells during the Awake State." *Nature* 440 (2006), 680–683.

Chapter 11: Our Takeaways

1. E. Turkheimer, "Three Laws of Behavior Genetics and What They Mean." *Current Directions in Psychological Science* 9 (2000), 160–164.
2. M. Francesconi, J. J. Heckman, "Child Development and Parental Investment: Introduction." *The Economic Journal* 126 (2016), F1–F27.

Chapter 12: Aging in Place

1. B. J. Cummings, E. Head, W. Ruehl, N. W. Milgram, C. W. Cotman, "The Canine as an Animal Model of Human Aging and Dementia." *Neurobiology of Aging* 17 (1996), 259–268.
2. M. M. Watowich, E. L. MacLean, B. Hare, J. Call, J. Kaminski, Á. Miklósi, N. Snyder-Mackler, "Age Influences Domestic Dog Cognitive Performance Independent of Average Breed Lifespan." *Animal Cognition* 23 (2020), 795–805.

Appendix III: Dealing with Diarrhea

1. C. A. Pugh, B. M. d. C. Bronsvoort, I. G. Handel, D. Querry, E. Rose, K. M. Summers, D. N. Clements, "Incidence Rates and Risk Factor Analyses for Owner Reported Vomiting and Diarrhoea in Labrador Retrievers—Findings from the Dogslife Cohort." *Preventive Veterinary Medicine* 140 (2017), 19–29.
2. D. P. Schlesinger, D. J. Joffe, "Raw Food Diets in Companion Animals: A Critical Review." *The Canadian Veterinary Journal* 52 (2011), 50; R. Finley, C. Ribble, J. Aramini, M. Vandermeer, M. Popa, M. Litman, R. Reid-Smith, "The Risk of Salmonellae Shedding by Dogs Fed Salmonella-contaminated Commercial Raw Food Diets." *The Canadian Veterinary Journal* 48 (2007), 69; O. Nilsson, "Hygiene Quality and Presence of ESBL-

Producing Escherichia coli in Raw Food Diets for Dogs." *Infection Ecology & Epidemiology* 5 (2015), 28758.

Notes

1. S.-E. Byosiere, P. A. Chouinard, T. J. Howell, P. C. Bennett, "What Do Dogs (Canis familiaris) See? A Review of Vision in Dogs and Implications for Cognition Research." *Psychonomic Bulletin & Review* 25 (2018), 1798–1813.
2. M. Siniscalchi, S. d'Ingeo, S. Fornelli, A. Quaranta, "Are Dogs Red–Green Colour Blind?" *Royal Society Open Science* 4 (2017), 170869.
3. Byosiere, Chouinard, Howell, Bennett, "What Do Dogs (Canis familiaris) See?" 1798–1813.
4. Byosiere, Chouinard, Howell, Bennett, "What Do Dogs (Canis familiaris) See?" 1798–1813.
5. A. Prichard, R. Chhibber, K. Athanassiades, V. Chiu, M. Spivak, G. S. Berns, "2D or Not 2D? An fMRI Study of How Dogs Visually Process Objects." *Animal Cognition* (2021), 1–9.
6. M. Makinodan, K. M. Rosen, S. Ito, G. Corfas, "A Critical Period for Social Experience–Dependent Oligodendrocyte Maturation and Myelination." *Science* 337 (2012), 1357–1360.
7. MacLean, E. L., B. Hare. (2018). "Enhanced Selection of Assistance and Explosive Detection Dogs Using Cognitive Measures." *Frontiers in Veterinary Science,* 5, 408876.
8. V. Wobber, B. Hare, J. Koler-Matznick, R. Wrangham, M. Tomasello, "Breed Differences in Domestic Dogs' (Canis familiaris) Comprehension of Human Communicative Signals." *Interaction Studies* 10 (2009), 206–224.
9. M. A. Udell, M. Ewald, N. R. Dorey, C. D. Wynne, "Exploring Breed Differences in Dogs (Canis familiaris): Does Exaggeration or Inhibition of Predatory Response Predict Performance on Human-Guided Tasks?" *Animal Behaviour* 89 (2014), 99–105.
10. P. Tomalski, M. H. Johnson, "The Effects of Early Adversity on the Adult and Developing Brain." *Current Opinion in Psychiatry* 23 (2010), 233–238.
11. S. K. Pal, "Parental Care in Free-Ranging Dogs, Canis familiaris." *Applied Animal Behaviour Science* 90 (2005), 31–47;

L. Boitani, P. Ciucci, A. Ortolani, "Behaviour and Social Ecology of Free-Ranging Dogs." *The Behavioural Biology of Dogs* (2007), 147–165; M. Paul, A. Bhadra, "The Great Indian Joint Families of Free-Ranging Dogs." *PloS One* 13 (2018), e0197328.

12. H. Vaterlaws-Whiteside, A. Hartmann, "Improving Puppy Behavior Using a New Standardized Socialization Program." *Applied Animal Behaviour Science* 197 (2017), 55–61.

13. P. Foyer, E. Wilsson, P. Jensen, "Levels of Maternal Care in Dogs Affect Adult Offspring Temperament." *Scientific Reports* 6 (2016), 19253.

14. G. Guardini, C. Mariti, J. Bowen, J. Fatjó, S. Ruzzante, A. Martorell, C. Sighieri, A. Gazzano, "Influence of Morning Maternal Care on the Behavioural Responses of 8-Week-Old Beagle Puppies to New Environmental and Social Stimuli." *Applied Animal Behaviour Science* 181 (2016), 137–144.

15. C. L. Battaglia, "Periods of Early Development and the Effects of Stimulation and Social Experiences in the Canine." *Journal of Veterinary Behavior* 4 (2009), 203–210.

16. J. W. Pilley, "Border Collie Comprehends Sentences Containing a Prepositional Object, Verb, and Direct Object." *Learning and Motivation* 44 (2013), 229–240; J. W. Pilley, A. K. Reid, "Border Collie Comprehends Object Names as Verbal Referents." *Behavioural Processes* 86 (2011), 184–195.

17. Adachi, I., Kuwahata, H., Fujita, K. (2007). "Dogs Recall Their Owner's Face Upon Hearing the Owner's Voice." *Animal Cognition,* 10, 17–21.

18. E. E. Bray, M. E. Gruen, G. E. Gnanadesikan, D. J. Horschler, K. M. Levy, B. S. Kennedy, B. A. Hare, E. L. MacLean, "Dog Cognitive Development: A Longitudinal Study across the First 2 Years of Life." *Animal Cognition* 24 (2021), 311–328.

INDEX

Page numbers of photographs and illustrations appear in *italics*.

cognition (*cont'd*):
Zindel's erratic scores on games, 97
See also: cooperative communication; executive function; memory; "physics"; self-control; *specific tests*

Congo, *vii*, xi–xiv, *xii*
aging and death of, *176, 178–181, 181*
brain capabilities and, 5
brain size and, 89
bringing the puppies home (2020) and, 130
as a Canine Companions service dog, xvi, 187
as an example for puppy parents, 15
as a great dog, xxv, 95, 181–82
hospital visits by, 120–21
as long-term memory study subject, 145–48, *147, 148,* 170
love of school, 84
playing cognition games with, 85
puppies mentored by, 146
service dog certification, 146
size of, xii, 85
skills of, xii–xiii, 15, 146
theory-of-mind ability in, 18
working at Puppy Kindergarten, xi–xii, xiii, 13–16, *16, 67,* 44–45, 176, 178

cooperative communication, xxiii, xxvi, 17–23, 19n, *20, 24,* 27–28, 32, 41–42, 83, 87
activities for raising a great dog, 167
Brian's dog Oreo and, 20–22
comparison of human infants, chimpanzees, and dogs, 23–24, *24*
Congo and, 18
dog breeds and, 89, 93
dogs communicating with each other, 23

dogs' understanding of human intention, 18, 20–22, 23, 27
domestication and when dogs became humanlike in communicating/cooperating, 24–28
emergence and mastery in puppies, 158–59, 160
example, Ashton, Aiden, and Dune play the shell game, 18–20, 19n, *20,* 28
example, wolf puppies play the shell game, 26, 27
human development of, 17–18, 27
as an inheritable trait, 91, 163, 163n, 226n75
multiple intelligence theory and, 31
pointing gesture and, 17–23, 88, 88–89, 158–59, *163,* 223n32
as predictive of success as a service dog and, 90, 93
as predictive of training success, 32, 46, 222n32
testing, 27–28, 34, 41
Wobber's comparison of wolflike to modern breeds, 224–26n75
See also pointing gesture

coprophagy. *See* poop-eating
Coren, Stanley, 83
crates. *See* dog crates
cylinder test, 46–48, *47, 49,* 161
Anya and Aries show self-restraint, 46–47
Weston's impulsive behavior, 47
Ying develops self-control, *47, 47,* 48–49, *49*

dachshund, 89
Darwin, Charles, 80, 145
detour game, 59–60
diarrhea, 211–16
discrimination (visual and affect) testing, in dogs, 32, *32*
dog beds, 131

Dog Cognition Longitudinal Battery
(DCLB), xxii, 33
dog crates
alone time in, 172
baby monitor and, 132
benefits of using, 133
as essential puppy supply, 130,
197
feeding inside the crate, 199
limiting time in, 133
sleeping in, 54, 132–33, 141
type used at Puppy Kindergarten,
130, 212
"Dog Emotion and Cognition" (free
online course), 167
doggy daycare, 173
Dognition, 84–90, 85n, 87n
"Canine Cognitive
Dimensions," 86
dataset on animal cognition, 87,
156
examining questions about the
cognitive performance of
different breeds, 87, 92
findings about dog brain size and
cognition, 89–90
the Gnanadesikans' research, 91,
91n
memory game, 89–90
number of dog owners and, 87
self-control game, 89–90
skills measured by, 86
what it is, 85–86
dogs, xxi
ability to cooperate and
communicate with humans,
xxiii, xxvi, 18–23, 19n, 20,
23–24, 24, 27–28, 32, 41–42,
83, 87
acting as alarm systems, 106
adult dogs performing better
on tests than puppies,
41–42
American Kennel Club
grouping, 88
brains of, 3–10

caring for the aging dog, 176–181
(*see also* senior dogs)
cultural shift, dogs as family
members, xx–xxi
domestication of, xxii, 27,
227n123
free-ranging (village or feral)
dogs, 227n123
the gifts they give us, 183
helplessness at birth, 5, 6–7
individuality in, 29–43
love of playing games, 84–85
"scoptopic vision" of, 219n7
sense of smell, 7
social intelligence of, xv, 12
social similarities to human
infants, 24
testing cognitive abilities of, 28,
31–43, 32, 32, 46, 84, 89–90,
221–23n32
theory of mind in, 18–24
vision and hearing of, at birth,
6–7
vision and perception of color,
219–220n7
when dogs became humanlike in
communicating/cooperating,
24–28
whiskers of, 7
wolf ancestors and, 25–26
See also puppy brain; *specific
breeds and traits*
dog toys
chew toys, 202
Kong, 203
recycled rubber balls, 202
stuffies, 202
dreams/dreaming, 151–53
do dogs dream, 151
memories in, 152–53
only recorded animal dream, 152
researchers with rats, 151–52

Ears Eyes Nose and Paws, xvin
English bulldog, 78
English sheepdogs, 96

cognition skills and, 222n32
consistency from puppy to
adult, 98
defined, 99
differences within a breed, 96–98
environment and, 99
great dogs and, 95, 100
individuality within litters, 140
research and recommending a
great dog, 98
Rainbow versus Sassy, 136–37,
137
Stanley's perfection, 138, *138*
temperament and, 98–113
what is best for a service dog, 98
Zindel as atypical of his breed,
96–98, 103
See also temperament
"physics" (or causality), *xx*, 34, 43,
66–72, 67n, *68*, 90
Arthur and retrieval skill, 77,
119–120
Congo and, 67
connectivity, 4, 67, 160
dog breeds and, 89
dogs' understanding of, 66–67,
67, 67n
emergence and mastery in
puppies, 160, 162
as fifth cognitive domain, 223n32
gravity and, 4, 66, 160
inheritability and, 91, 164
lesson that puppies are not
Einstein (they are not good at
it), 168
physics games, 34, 67, *68*,
68–69, 72, 89, 223n32
success as a service dog and, *xx*,
69, 90
Wisdom's skills, 119–20
Zax and the physical principle of
connectivity, 67
Zax and the physics game, 67–69
pointing gesture, 17–23, 88–89
Aidan and Dune follow pointing
gestures, 20

dog breeds and, 88, *88*
emergence and mastery in
puppies, 158–59, *163*
games for testing in dogs, 223n32
Oreo and, 21–22
as predictive of service and
military dogs training success,
xx, 222n32
See also cooperative
communication
poodles, 83, 89
Moustache and French troops,
106
toy poodles, Woober's research
and, 225n75
poop-eating (coprophagy), xiv,
64–66
graduation as service dogs likely
and, 65, 66, 72
a liability that can be an asset, 66
trainability of a dog and, 66
positive reinforcement method,
51–52
puppy brain, xxvi, 3–10
affected by experiences, 8–9 (*see
also* environment)
at birth, 6
cat brain versus, 5
cognition and, 6
compared to other mammals, 8
corpus callosum, 8
differences related to breed
size, 89
Dog Cognition Longitudinal
Battery (DCLB) and, xxii, *33*
at eight weeks old, 9–10
emergence of cognitive profile, 10
emergence of temperament, 9–10
executive function (problem
solving), 48, 89
growth of, xxii, 3–9
gyrification (growth of cortical
folds), 8
hearing and, 7, 157
how the puppy intelligence
develops, 156–57

testing for communicative skills, 32, 34
See also cooperative communication
socialization and training of puppies
adoption at around eight weeks of age and, 117
age to promote positive interactions, 127
age to start each skill, 52–53
basic life skills (commands) and, xxv, xxvn
cognitive games associated with training success, 229–230n161
cognitive traits for, xxiii
comparison of puppies raised by volunteers and puppies raised at the kindergarten, 117–123
a day in the life of Wisdom, 118–123, *120, 122*
detour game and, 59–60
effects of early life experiences, 116, 118–123, 164, 226n16, 228n126
effects of mothering style and littermates, 123–27, *124, 125,* 227–28n125, 227n123
every puppy is different, 142
final exams, 53
goal for skills at twenty weeks, 211
Home Schedule, 206–11
impact of mistreatment or neglect, 123
inducing fatigue as a tool for, 55–57, 60
lessons for raising a great dog and, 170–72, *171*
midterm testing, 53
for military doors, 59–60
positive experiences for, 121
positive reinforcement method, 51–52
puppy classes or puppy school, 127
self-control and, 51
sleep training, 135–142, *138, 139*

study of intensive socialization or "gentling," 126–27, 228n126
teaching manners, xxvi, 51–52
threshold for, 123, 164
training sheet of skills, 52–53
treats used for, 51–52
window for puppies, xxiii, 9, 116, 117, 164, 172, 220–21n9
See also raising a great dog; *specific skills*
squirrel monkey, 48
stranger danger test, *104,* 104–6
Arthur meets a stranger, 104–5
Zindel meets a stranger, 105–6
Swedish Armed Forces, 228n125

teeth
appearance of permanent teeth at twelve weeks, energy burst, and chewing, 50–51
milk teeth, 50
teething toys, 51
Weston's chewing and misbehavior, 54
temperament, 98–104
Arthur's response to tests, 100–102, *101, 102,* 104–5
breeding for, 111–12
Canine Companions selection for friendliness, 104, 106, 111
C-BARQ surveys and, 65n
classic Labrador, 104
classic temperament test: meeting something new (robots), 100–104, *101, 102, 103*
cognition and, 99
as consistent through time, 98
defined, 99
easy or nonreactive dogs, 102, 103, 111
emergence in puppies, 9–10, 100
emotional reactivity and, xviii, 103
exceptionally calm and friendly: Arthur, Zindel, and Westley, 111

About the Authors

BRIAN HARE is a professor in the Department of Evolutionary Anthropology and the Center for Cognitive Neuroscience at Duke University, where he founded the Duke Canine Cognition Center. VANESSA WOODS is a research scientist at the center as well as an award-winning journalist and the author of *Bonobo Handshake*. Hare and Woods are married and live in North Carolina. They are the authors of *Survival of the Friendliest* and *The Genius of Dogs*.